Cross Currents

a coming-of-age novella

Bk 34

by

George G. Pinneo

Begun 5/31/2018

Current Date: 9/9/2018

Overgaard Arizona

Author's Note: this novella is a work of *fiction*; names, characters and incidents are products of the Author's imagination.

All Rights Reserved:
No part of this book may be reproduced, stored in a retrieval system or transmitted by any means: electronic, mechanical, photocopying, recording or otherwise, without written permission from the **Author**.

Realistic science fiction novels by George G. Pinneo

The Bergmann Series

Bergmann's Commitment - Batwing, 4th Ed. Bk 1

Forming Bergmann's Team, 4th Ed. Bk 2

Driving Bergmann's Comet, 4th Ed. Bk 3

Bergmann's Equestrian Venture, 4th Ed. Bk 4

Bergmann's Orbitat, 4th Ed. Bk 5

Bergmann's Interplanetary Venture, 4th Ed. Bk 6

Bergmann's Martian Venture, 4th Ed. Bk 7

Bergmann's Martian Colony, 4th Ed. Bk 8

Sunrise Venture, 4th Ed. Bk 9

The *Planet Scout Series*

Factor - Threshold to Space, 4th Ed. Bk 10

Factor - On to Mars, 4th Ed. Bk 11

Planet Scout - Raptor's World, 3rd Ed. Bk 12

Planet Scout – Birenso, 3rd Ed. Bk 13

Planet Scout – Timoron, 3rd Ed. Bk 14

Timoron Colony, 3rd Ed. Bk 15

Planet Scout – Brandywine, 3rd Ed. BK 16

Planet Scout – Polychrome, 2nd Ed. Bk 17

Planet Scout – Pacifica, 1st Ed. Bk 18

Planet Scout – Kavalor, 1st Ed. Bk 19

Planet Scout – Vondalar, 1st Ed. Bk 20

Technomilitary Mystery - Novellas
The RPVs At War series

New Beginning 1st Ed. Bk 21
ISBN-13: 978-1542346726

Gunner in the Sky 1st Ed. Bk 22
ISBN-13: 978-1542382007

Rifle in the Sky 1st Ed. Bk 23
ISBN-13: 978-1542714754

Sniperbird 1st Ed. Bk 26
ISBN-13: 978-1548901967

Slinker 1st Ed. Bk 29
ISBN-13: 978-1984345042

Smugglers in the Dark 1st Ed. Bk 30
ISBN-13: 978-1986038447

Hard Science Fiction-Novellas

Seaborne 1st Ed. Bk 24
ISBN-13: 978-1546833796

Plesiosaur 1st Ed. Bk 25
ISBN-13: 978-1548508296

In New Ground 1st Ed. Bk 28
ISBN-13: 978-1983746734

New Arrangements 1st Ed. Bk 32
ISBN-13: 978-1718802971

Novellas on Aging

Alone 1st Ed. Bk 31
ISBN-13: 978-1986713283

Moving On, Alone 1st Ed. Bk 33
ISBN-13: 978-1720600480

Coming-of-Age Novels

Making It 1st Ed. Bk 27
ISBN-13: 978-1981950881

Cross Currents, 1st Ed. Bk 34
ISBN-13:

All paperback books are available from Amazon. All books are also available as Kindle e-Books.

Table of Contents

Chapter 1	New Opportunity	Page 8
Chapter 2	Looking Around	Page 29
Chapter 3	Offer	Page 46
Chapter 4	New Jobs	Page 63
Chapter 5	Going Home	Page 79
Chapter 6	Dinner	Page 95
Chapter 7	Heading West	Page 106
Chapter 8	Reset	Page 121
Chapter 9	Southern California	Page 135
Chapter 10	Starting Work	Page 138
Chapter 11	Designing an Experiment	Page 157
Chapter 12	Low Noise Amplifier	Page 174
Chapter 13	Super Polish	Page 200
Chapter 14	Improved LNA	Page 217
Chapter 15	New Digs	Page 219
Chapter 16	New Relationship	Page 231
Chapter 17	Substrates	Page 245
The End		Page 249

Cast of Characters   Page 250
About the Author   Page 252

Chapter 1   New Opportunity

Grant Porter considered the computer display in front of him. He was on an interview-trip to Southern California; this was the first step in applying for a job at Northrop Grumman in Redondo Beach, California, a totally new place for him. NG was a major aerospace firm that built high frequency, high power satellite electronics, among many other products.

The last section of NG's electronic job application wanted him to summarize what originality *he* might bring to their organization. He was a newly minted engineer from Purdue whose diploma was still wet with ink! He wondered how specific to be? He was 21; his diploma said he was now a B. S. Materials Engineer. He had some chops in metal and glass casting; *jewelry*, really.

Working as an apprentice jeweler had allowed him to earn *some* of the cost of his Bachelor's expenses. An insurance policy his Dad had taken out years ago had helped pay much of the cost. He'd gotten a job in the university machine shop as a junior-level machinist that had paid some of his expenses and allowed him to learn more of machining common metals for fixtures and test articles that post-graduates and professors needed made.

So he owed something like $11,000 at graduation; while that seemed like a huge debt, it was almost trivial compared to the loans some of his classmates had accrued! He played with the wording on the electronic document to highlight his skills in casting silver, bronze and gold rings and pendants, emphasizing the design requirements of both 'Delft Clay' and 'lost-wax' investment casting procedures. He typed a short paragraph detailing his experience at hand lay-up of epoxy-glass and epoxy-graphite aircraft skins. He didn't mention his woodworking skills, concerned they might sound too much like a high school shop class.

He had no idea how the HR people would read this last section of his application! He played with the wording of how he'd taught himself to slump colored 'art' glass into decorative dishes and how to fuse glass powders into colorful, light-catching pendants.

The 9-page application form had taken him almost 80 minutes to complete. He glanced up to see that Mrs. Miller, a 'matronly woman in maybe her late fifties, was eyeing him from across the room that held 'computer carrels' in the HR offices of Space Park, Redondo Beach. Was she telling him he'd spent enough time on the form?

He quickly added a section on torch-soldering: using silver and gold hard solders to attach pendants to chains and dangles to bracelets. He then reread this last section for typos and sighed; the form was as complete as he thought he could make it. He wanted them to understand he was a *hands-on* type, not just a book-reader and memorizer! He hadn't found a box to check for that!

He sighed, clicked 'Save' and sat more erect, hoping that there was some value to this very complex corporation in what he'd just written. He stood up and stretched as Mrs. Miller walked towards him.

"Are you comfortable with what you've captured for us, Grant?" Mrs. Miller asked.

He smiled. "I am; I'm concerned I might have been too specific about my personal skills. I don't know what you expect to see in such an application. I brought a photo album of some of my better jewelry; I don't know if that's something you want to see or not." He tried to read her face but could not.

"I'm going to read the whole form, Grant. You might want to walk over to Building S, the next building to the West; they have an espresso cart out front. Our big cafeteria is in the basement; you might find pretty much anything you want, there! I suppose it's a bit early for lunch, actually," she replied. "I want to talk to you, but give me half an hour or so to read through this, please."

He smiled: "espresso sounds very good, Ma'am! I'll go check out the espresso and then come back here!" He exited the R4 building past two other guys waiting to talk to HR, undoubtedly applicants, and could see the colorful striped canvas roof of a small, 'one-barista' java cart on the sidewalk ahead of him.

He knew he had to 'pass' the application process before he would be interviewed. If they didn't like the *application*, he probably wouldn't *be* interviewed!

He wondered what the dress code was at NG?He noted some suits on older men walking by; he'd worn just a long-sleeved button-down pin-point cotton shirt and chinos; he carried a wool sport coat, but it was warm enough he didn't want to put it on. He had no clue to what was *underdressed* out here, but he might be a bit casual!

He hoped the interviewers would try to size him up for sincerity, attitude, good sense and entry-level technical skills; the application for a job as a starting engineer, fresh out of engine school, was just the beginning. He was guessing any interview would start after lunch, if they liked his application.

The doppio espresso machiatto was hot; it was a bit mild for him, but it was eminently drinkable, once he'd put some sweetener in it. He wandered around the steps leading down to a large cafeteria and perused the lunch menu posted there. He decided he might eat lunches here, if they made him an acceptable offer.

He was aware that he felt *suspended* in a new place: he was graduated, but had no notion of what his life would be next! He wanted to begin *something* in this new part of his life!

He'd taken 19 corporate interviews so far; most had been with interview teams on campus: a very preliminary kind of thing; 6 of them had resulted in short 3-day trips to various firms in several Mid-Western cities. This was the only transcontinental interview trip he taken so far and his first chance to visit the Golden State.

He *had* a first offer at a firm just outside Cleveland that liked his machining skills; they made a variety of smallish, complex investment castings in stainless steels. He would start learning CAD designing immediately if he went with them; he might live at home or at least go home to do laundry if he took that job offer! He had mixed feelings about starting down that track; once you were 'indoctrinated' into a specific type of CAD, it gave you a serious job but limited what you actually might do in other disciplines.

One of the jobs he was interested in was in a solar power firm in Indianapolis; they were going to combine a network of solar power systems with electric vehicle charging: a potentially serendipitous combination. It was a entrepreneurial thing with all the risks and rewards of a new concept. He had no idea of whether it would be a good thing for *him*. His Dad thought the process of learning how to build such a firm would teach him things quickly. He couldn't commit to anything in that start-up yet! The feeling of being suspended somehow, between his life as a student and that of a working engineer *nagged* at him!

Another promising job was up in a Boston suburb in microelectronics. That was a promising offer in terms of probable growth, but he worried about the costs of living and working in 'Taxachusetts'.

NG was his sole interview in aerospace and on the West Coast; the cost of living out here might be a bit higher than anything he knew of. He knew that California had a *lot* of firms working in that general bailiwick of 'space and electronics' development; he had done a quick sort of those firms into vehicle structures and electronics but hadn't refined that sorting into any more detail. He might do structures, but was leaning toward electronics because of the many sub-specialities and the rapid pace of microelectronic technology development.

He looked across the intersection of Marine Avenue and Aviation Boulevard as steady streams of cars moved in 4 directions: north, south, east and west, in the sunny California weather that seemed like spring to him! Southern California traffic was a *lot* busier than what he saw back in Indiana and Ohio!

His interview package had him staying in a Residence Inn in the next town west along the Pacific coast: Manhattan Beach, which seemed a non-sequitur of a name. He was going to walk down and see the Pacific Ocean up close at some time before he flew back to the Midwest.

He guessed he'd given Mrs. Miller enough time to scan through his application and very short resume. He walked back to the R1 lobby, trying not to look anxious.

An attractive young black woman, he couldn't begin to guess her age, beckoned him as she read his temporary visitor's badge. "I'm Amy Borden, please follow me; we're going to a different office where you'll meet with Mrs. Miller and our Mr. Robert McPage; he's one of our technical managers. He liked something in your application!" she said as she led him upstairs to the second floor.

That was maybe a *good* sign, he thought to himself! I'll bet they don't get many skilled jewelers coming in to apply for engineering jobs!

Amy turned him over to Mrs. Miller and Robert McPage. McPage was a stout, older man with gray sideburns, who studied him through glasses as they shook hands. "So you're a jeweler?" McPage began.

"Well... *apprentice* jeweler, maybe," he said with a smile. "I can design and copy pendants, rings, bracelets in silver, bronze and gold, and I can design and make some simple glass pendants and dishes. I can assemble jewelry pieces using torch-soldering of silver and gold hard solder." Grant paused to see if that would satisfy McPage.

"Tell me why you'd use one metal over another, please," McPage asked, trying to gage what he knew.

Grant took a quick breath. "Most women buy silver over bronze although it costs a bit more. Some women have the means to buy gold, but in northwest Ohio, not a lot of people buy gold jewelry. Bronze is about $14 per pound; silver is about $19 per troy ounce and gold is about $1,390 per troy ounce. In Ohio at least, there's a price point for silver pendants in the... say $60-$75 range. Most gold pendants of any size are going to go... maybe $125 to $175. We sell a lot more Sterling silver jewelry, 92.5 percent silver plus 7.5 percent copper, than gold. What a jeweler really likes is to have someone come in and ask for something we already have in silver, custom cast in gold. Those people will spend more to have a piece in 14 karat yellow gold, 58 percent gold, 25 percent silver and 17 percent copper," Grant replied, hoping McPage really wanted that level of detail.

"Which metal would you rather cast, Grant?" McPage asked.

"Well, gold doesn't oxidize; *I* cast 14 Karat gold at about 1,500 degrees F. I cast Sterling silver at about 1,700 degrees F and I cast white bronze, a leaded bronze, at about 1,860 degrees F. So, gold is easiest and more fluid than either silver or bronze. Silver would be my second choice because it doesn't oxidize very much if I use a graphite crucible in my electromelter. Bronze oxidizes a little faster than silver, but the leaded bronze I use is designed to be pretty fluid, so I get good dense castings all the time. Younger people, mostly women, of course, can afford a *lot* of white bronze jewelry!" Grant explained with a small smile.

"Into what do you cast these pendants?" McPage asked, studying Grant's eyes.

This guy knows some jewelry, Grant realized. "I use, we use, two techniques: the simplest is what is called 'Delft Clay' casting, a subset of sand-casting. The more precise and more time-consuming procedure is the traditional 'lost-wax investment casting' that has been around for almost 5,000 years. Investment casting is much more precise and allows a much wider range of detail, but it takes longer to prepare the model and the investment. I can walk into my shop and make a Delft Clay 'sand dollar', a small dolphin or seahorse in about an hour. If I cast a detailed, *hollow* pendant with fine cross-sections in wax, it'll take me something like 8 or 9 hours to complete, regardless of the metal I use," Grant explained.

McPage nodded and smiled. "Very good explanation, Grant; I have a brother whose wife owns and operates a small jewelry shop just west of us over in Manhattan Beach up the street from *The Kettle*, a restaurant I recommend to you. Do you have a *specialization* you would like to consider here at NG?"

"I do not, Mr. McPage; I know some metallurgy; I know *something* of hand-layup of epoxy/fiber composites and epoxies in general; I know something of hard soldering. I would like to learn more about microelectronics because it has become so critical in aerospace. I would guess that your business might accommodate all three areas," Grant replied, hoping he'd said *something* that was meaningful to this man he didn't know.

"Very good answer, Grant! We are going to go to lunch with some other potential new hires. There are 8 of you new grads out here this week; the 8 of you might constitute a 'summer' new-hire class, if we can come to some mutual arrangements! Leona and I are going to walk you over to a corner of the cafeteria where we can continue our conversation over lunch!" McPage replied.

OK! The two HR people were going to stick a bunch of greenies in the same space and see how we interact. That might be interesting, he thought!

McPage gestured Miller should lead the way. Amy brought a row of 3 young women and 4 young men out of a waiting lounge and they went out of the building with Miller leading and McPage trailing. Grant tried to scan the 7 other greenies as they walked. One of the guys wore a 3-piece suit of dark blue wool; he was overdressed in comparison! The other two guys were dressed as he was: fairly casual. The 3 young women wore a pants suit and 2 dark 'business dresses', if that's what they were; he couldn't see more because they were all behind him where he followed Miller.

A few minutes later they were in one of the serving lines in the basement cafeteria. He opted for a hamburger with pickles and tomato slices on a toasted bun; it came with fries, so he only had to fill a glass with ice and Diet Coke and carried both items on a tray toward where Amy indicated they would sit.

The cafeteria offered several 'meals' prepared in stainless steam trays: something chicken with pasta; some pork loin thing with gravy and some kind of Salisbury steak dish with broccoli. This might be a good place for a leisurely lunch if he opted to not just go to the McD's across the Aviation Boulevard in Manhattan Beach for a quick, light lunch.

When they joined Mrs. Miller at an empty table, the girl, woman, in the pants suit, took the chair opposite him. She was pretty with short, dark brown hair curled against her head; she nodded to him and he nodded back. She had very vivid dark eyes. He took a bite of the burger and found it was pretty good for 'slow food'.

"Amanda McCormick, this is Grant Porter. Amanda is a ceramics engineer from Rensselaer; Grant is a materials engineer from Purdue," Miller said, introducing them. "Grant is a part-time jeweler, so he has some grounding in the casting of metals." They shook hands across the table.

Amanda's very neat eyebrows went up. "That's got to be interesting work! You cast rings and pendants?" she asked.

"I do; I cast mostly silver and bronze; fewer people can afford gold, although all 3 metals have been cast from antiquity!" Grant replied with a head-bob after he put the burger down and used a napkin. This young woman was only a handful he'd met at Perdue. There were more women in engineering, but they were only a few.

The young Asian woman beside Amanda said: "I'm My-An Nguyen; my Mother worked in a very small jewelry shop in Saigon before she moved us over here. She cast mostly gold and some silver, Sterling, of course! I got to watch from across the room so I wouldn't get splashed with molten metal!" she said with a smile.

"I've not been to the Far East; I'm sure the metallurgy is very similar! Back in Ohio in my hometown, most people buy Sterling silver pendants, rings and bracelets, because they are more affordable; only a few can afford larger pieces in gold, usually 14 Karat yellow gold," Grant replied.

"Did you ever do anything in platinum? My Mother *hated* to work that metal: it has a high melting point?" Nguyen asked, eyebrows up.

"Roger *that*! Platinum has a *very* high melting point; it requires something like 3,300 degrees F! Yellow gold is about 1,500 degrees and Sterling silver is about 1,850 degrees. I've heard some casters complain about having to use oxyacetylene torches to melt platinum. I like to use a small electromelter for whatever I'm casting; the graphite crucible helps suppress oxidation. An electromelter is effortless to use, at least on small castings," Grant replied, hoping he wasn't just blathering.

"You brought a picture-book out, Grant?" McPage asked from where he sat on the other side of the table from Nguyen.

"Ah, I did; it's in my iPad. I, ah..." Grant unbuckled his belt and unsnapped the buckle. "*This* is white bronze I cast in Delft Clay; I cast it flat and then very carefully curved it." He handed the belt buckle across the table to McPage. McPage studied it for a minute and handed it to Nguyen.

"The white bronze casting chunks I buy are something like 60 percent copper with some tin and a small amount of lead. The company I buy it from doesn't really specify the exact mix. The lead is a 'low melter' and helps increase fluidity. It's very strong; I've been wearing this buckle for almost 4 years now!" Grant declared.

Grant gestured Nguyen should pass it around the table. He'd just not be able to stand up until he got it back!

When McCormick got it, she studied it and flipped the bail back and forth, smiling. "Pretty simple, strong design: 'GP' joined together!"

He nodded; her eyes were striking! She passed it to Gannon, the over-dressed guy, who looked a little warm in the air-conditioned basement with a coat, tie and vest.

When the buckle came back to him he quickly snapped it back on his leather belt.

The table talk was all small talk until Mrs. Miller led them back to R1. Grant ducked into a restroom before joining the others in a small, glass-walled conference room. He made sure there was no food on his face or in his front teeth, a bit nervous about what would come next.

He took a chair at the table and placed his leather iPad cover on the table on top of his leather folio; he had a couple copies of his single-page resume, which included his part time work experience and a reference sheet on precious metal casting.

Amanda McCormick took the chair across from him, flashing a very nice smile as she placed an iPad on the table in front of her. When all 8 interviewees were back in the room, Amy Borden came in with a tablet; she prepared to begin taking notes.

Mrs. Miller used a small laptop and called up a document she referred to by leaning in to see it better.

"Welcome to Northrop Grumman Space Park, Redondo Beach. What we're going to do next is have you introduce yourself to us. Take your time; it might be interesting to have you mention your hobbies," she said, smiling.

"Let's start with My-An since she's on the end," and she nodded to the small Asian woman.
"I am My-An Nguyen; I am a Vietnamese emigre to Los Angeles. I grew up in Orange County and now have a BS in mechanical engineering from the University of Rochester in New York. I like to read of Asian history. I run, mornings usually." She bobbed her head that she was finished, almost shyly.
"I am Amanda McCormick from Sandusky Ohio; my bachelor's degree is in Ceramic Engineering from Rensselaer University in New York. I'm a *potter*; I like working clay; I've done some glass work, but I don't have room for a kiln at home. I'm enjoying this trip to Southern California where the weather is amazingly comfortable. Thank you!" She flashed another smile at Grant. Grant thought: Ohio girl! How ironic to meet her out here!
"I'm Keith Gannon, Industrial Engineering from the University of Southern California in Los Angeles. I'm working towards a master's in Engineering Management in my spare time. I play rugby on weekends!" Keith's smile was big but Grant saw it as forced.
The other four interviewees spoke in turn; one was a woodworker; one was a bowler, one was a novice writer and one was a fisherman. Grant was last.
"I'm Grant Porter with a BS in Materials Engineering from Purdue. As I said at lunch, jewelry is both a hobby and a part time job.

Jewelry is an *ancient* practice; both the engineer and the historian in me are intrigued with how creative the jewelers of the Middle East were with *none* of the tools we have available today! There are an abundance of silver, bronze and gold artifacts, likely made with beeswax models, from as far back as 2,500 BC, or BCE, if you're into ancient history!"

Grant watched McPage's face. "I do some reading, some woodworking and some camping and hiking. I particularly like canoe camping!" Grant nodded and quit. He didn't want to seem overbearing. He left out that he was a student pilot; they didn't need to know that.

"Very good, now we know a little more about you," McPage said, making some notes on his iPad. "Our next sessions will be to spend a little more time with each of you individually. You know from the interview packet that you will have to pass a drug test and a medical exam; you know that any criminal record will likely prevent us from making you an offer." He studied them all briefly as he said that. The computer application form had taken an image of their drivers' licenses; that appeared to be the source of most of the information NG would be able to pull up about them from public records.

"What takes the longest time for new hires is to get you a *preliminary* security clearance. We need that to be comfortable you can work on the wide range of technology we build here and in affiliated facilities across the state and the country. *Everything* we do here is either classified or proprietary. We'll talk about that from time to time," McPage explained.

"If you've been *out* of the country; the security investigation can take a little longer, sometimes more than a month. What we can do in the meantime, is offer a *temporary clearance subject to further review.* Sometimes we simply can't find any reason that you *shouldn't* be cleared, but sometimes there are issues. How many of you have been out of the country? Grant held his hand up with both McCormick, Gannon and the other 3 men.

"My-An, your emigration will need some study; your application said you're a naturalized citizen. That's good! Your Mother is here?" McPage asked. My-An nodded. "She is: we live in Little Saigon down in Orange County! My three sisters all live at home with her," Nguyen responded. "She has our passports, Sir."

McPage nodded. "Can you bring it in tomorrow, please?" he asked My-An. "Now, *who's* been to Europe?" McPage asked. McCormick, Gannon and the other 3 men all raised hands. McPage smiled. "Where'd *you* go, Grant?" he asked.

"I went canoe camping up in Ontario, in Algonquin Provincial Park, actually," Grant said, smiling.

"Ontario is pretty easy; Europe is a bit more involved. We didn't ask to see your passports, but if you can bring those to us in the next week or so, that may help us get an interim security classification. I'd like to start with Grant. The seven of you can go outside with Mrs. Miller and we'll get started." McPage was still smiling.

Once the door closed behind the others, McPage asked: "so, what *didn't* you tell me about your interests, Grant?"

Grant hesitated for a long moment. "Well, I like to build things; my Dad and I do a lot of... little projects in the garage-shop. I like to see if I can cast... unusual shapes." He paused; he didn't know what McPage wanted to hear.

"Oh, I'm a student pilot. One of my Dad's friends is the jeweler I worked for. He has a Cessna 172; I've got a whole 42 hours in it! I've soloed and... when I have more money, I'll take the time to get current and then take the FAA exam for a private pilot certificate." He smiled and paused, not at all sure what McPage wanted to hear.

"I hunt pheasant and some ducks with my Dad and his friends sometimes in the Fall." He thought for a moment. "I'm not sure there's anything else to say, Mr. McPage."

McPage smiled and nodded. "A pilot wannabe; that's interesting! It takes some skill to fly a plane carefully. I always *wanted* to do that, but I never set the money aside and did it. Do it as soon as you can, Grant; we learn and retain more rapidly when we are young!"

He smiled and shook his head. "If you like our offer, you will probably learn a number of new skills here at NG, Grant! I'm done with you for now. I would like to see your passport or at least get the number; can you call your parents to get the number?" McPage asked.

"Of course I can! I'll call them this evening; I could get you the passport number tomorrow morning, I guess," he replied. Apparently, they were going to make him an offer! That was an unexpected relief!

A minute later, McPage brought McCormick into the room as Grant took a seat outside on a padded vinyl chair beside the others. Amy walked up to him from an office where Mrs. Miller sat in front of a big desktop computer. "Mrs. Miller asked me to give you this informational packet, Grant." She handed him an 8-inch by 12-inch manila envelope.

"It contains a list of local rental housing and some local eateries and has a couple of local maps. We *will* be making you an offer, Grant! We'll detail the offer to you in the morning. Be back here at 8 AM, please. The list of rental properties is scattered out through the LA Basin, from here to Corona and Riverside to the east and up to West Covina in the northeast, as well as down to Anaheim in the south. The region to the north, 'downtown L. A.', gets almost as pricey as the beach cities, but one *might* cruise a neighborhood looking for 'room-for-rent' signs. We might assist you with a guarantee of employment for some home owners in some cases," Amy said.

"The *easiest* thing for you to do, is to consider some of what I'll call 'singles apartments'; there's a short list of relatively new, smaller studios and apartments. We see a lot of our new hires go to the region out along the 91 Freeway: Cerritos, Buena Park, Fullerton, Orange, Yorba Linda and Corona. There have been some apartment complexes co-located with chain restaurants along the 91 out that way. Some of them are larger than others, but all offer compact, relatively new apartments. Some of them are quite modern and quiet."

"Commuting on the 91 Freeway is a relatively easy commute, generally speaking. Engineers at NG don't clock in, in any formal sense; you'll keep your own hours and log them on a computer; we allow some flexing of start and stop times. We understand that new engineers may choose to keep their own hours, subject to your immediate superior's approval, of course. Your assigned starting position will allow you to arrange your schedule with your new supervisor, to some extent."

"Mrs. Miller will explain how we induct new graduates in the session tomorrow morning. In general, we will assign new hires to one of several departments, based on their skills and interests. You'll be assigned to a department manager who will use you to solve some of his or her problems. Go check out some places you might rent or lease; leasing might save you some money. Most property operators will likely ask for first month, last month and current month in advance," Amy explained. "Here's my card if you have more questions!" She smiled and made a shooing gesture.

Chapter 2   Looking Around

Grant chuckled, nodded and stood up. "Thank you, Amy! I'll see you in the morning with my passport number." He would drive his little rental Toyota out the 91 and look around; he had *no* idea how to pick a place to live out here! He could just look for now and then maybe look more intensively if he liked the offer. It would be good to just look around in this new place! He wondered if he could ship his few books, desktop comp and clothes out by air? He *could* drive them out in the hand-me-down pickup he'd gotten from his Father; there *was* a certain romantic notion to driving out here across the States!

He wondered if the ten-year old pickup could *make* a 2,200 mile trip without breaking down? He knew one front tire was bald and would need to be replaced; he might even buy two and put them on the front end as insurance. He and his Dad could change the oil filter and oil easily enough. He might want to have the brakes checked. He'd want the truck out here to commute into work every day; there might be car-pools, but he'd be more comfortable being a little bit independent of other people's schedules.

He opened the new packet and studied the maps, adding them to the map he'd picked up at the Enterprise rental desk at the airport, which was very limited in range and detail. He pulled into a 7/11 and got a diet cola and got back in the tiny rental car.

The 91 Freeway wasn't too bad at 3 in the afternoon; he guessed coming back in to the Residence Inn in Manhattan Beach would be against the traffic flow if he waited until 6 PM or so. The 91 traffic was a broad, noisy river of steel and plastics, 4 and 5 lanes each way, flowing east and west at 60 plus miles per hour! Cleveland Ohio had some highways this big and this busy!

He looked up two apartment complexes: "Tesseract Towers' in Yorba Linda and 'New Rameno Village' in Orange. He was nervous as he drove east on the 91 Freeway: this was a major highway! He wondered how bad it was at rush hour?

He pulled into a parking lot in front of New Rameno Village; the city of Orange was south of the 91, off to the east of the 55 Freeway. Suburbia stretched in all directions, 'Village' might be appropriate; the complex was four big 10-story buildings arranged in a big rectangle with a bunch of smallish restaurants arranged around the parking lot within easy walking distance of the apartments.

He found the sales office and went inside. He studied some displays showing views of the complex taken from the air, with major roads detailed. The 57 and the 55 went north and south. The 57 connected to US 5 which ran northwest towards downtown LA; the US 405 paralleled the Pacific Coast Highway or 'PCH'. PCH ran all the way up past Santa Monica, a locale too rich for his blood! He'd listened to his Dad explain that *all* of the beach cities were too expensive for his budget; Dad had mentioned El Monte and Walnut as places that might be more moderately priced. Amy had aimed him at the 91 'corridor' for an easy commute.

He looked at three model apartments: *medium small* with two beds, *smaller* with one bed and *small* with the sleeping 'compartment' off the very small living room/kitchen. How much room did he need? The dorms at West Lafayette, Indiana, were very plain, but weren't cramped; he did a lot of studying in the library because it was quiet. He might be fine in either the smaller or small 'studio'.

He asked a slight Asian woman of indeterminable age to see an available unit. She dug up a key and led him up an elevator to the 4th floor. The unit seemed very clean; the furniture was hardly used at all. The kitchen, where he would likely eat only breakfast, was clean and modern; the kitchen sink was clean and shiny. This wasn't a bad place!

"This unit is $775 per month, Sir: 3 months down at signing. I have a list of the restaurants that surround our big parking lot: they include a McDonalds, an Arby's, an Outback Steak House and a breakfast place called 'Farmers'. Farmers has very good breakfasts, especially omelets and croissants!" she said with a smile. "There's a Texas steak house just 2 blocks north and a Mulligan's Irish Pub one block east. There's a Coffee Bean and Tea Leaf, too."

Grant asked if she could show him the middle-sized model? A few minutes later she showed him a unit on the 8th floor, which had a nice view out to the east. The second unit was as clean as the first one. This was marginally larger but quite modern. "Thank you, I'll probably come back after I think about it. I don't need a place until maybe next week; will that work?" he asked.

The woman was guarded but was reassuring that they would have several units available through the weekend. "Its better to make arrangements as early as you can, then there's less chance someone will come in and sign a lease before you return!"

A few minutes later, Grant went back out to his rental car with a handful of brochures and pricing sheets. He drove back to the 91 and went east. Tesseract Towers was more impressive: four 20-story towers arranged in a square around a modern bar overlooking several very green golf courses to the North in Chino Hills. The view from an 18th floor unit was very nice! The studio apartment was clean and as modern as the one at New Rameno. The 'TT' units were priced within 15 dollars of those at New Rameno Village. He guessed there must be some pressure to be competitive.

He was oddly tired, maybe jet-lagged; he decided he'd have to come back when he was more alert. He thanked her and went directly from the sales office out to his little car. He stopped at a Starbucks and got a doppio espresso machiatto; that refreshed him enough as he drove west through lighter traffic toward the setting sun out over the Pacific. There was beautifully clear weather out here!

He went into the Residence Inn lobby and asked about local restaurants. The young woman clerk recommended he walk down toward the Manhattan Beach Pier. "There are a bunch of good restaurants down there: The Kettle, Mama D's and Darren's for seafood. There's a chocolate shop and and ice-cream place and a Starbucks, of course. Be sure to check-out the Strand; it runs up into Marina Del Rey if you like strolling!" she suggested.

He nodded and decided he'd drive partway down PCH toward Manhattan Beach Boulevard. There might be some street parking free after 6 PM, per one brochure. He took the first empty parking space after putting a quarter in the meter, and carrying his sports coat against the cool air coming up the street off the ocean, walked on west! He found The Kettle across the street from a Starbucks and scanned the menu board. He could eat here! He went across the street and got another doppio espresso machiatto and then headed on down to the Manhattan Beach Pier, looking all around at what to him, seemed very different and strange!

There was an aquarium out on the end of the pier, closed now. The view to the north revealed the hills of Malibu beyond Santa Monica: lights coming on as the hills hid the setting sun. The view was like something out of a movie! There were no *views* like this back in Ohio!

He was jarred by how many young women wore bikinis! *Damn*! Some of them were too young to care, but some were very much into showing a lot of skin! Some walked up from the sandy beach; some skated, some skate-boarded and some biked the Strand, north or south! There were a *lot* of people out walking some with dogs, some with children in wagons, some in strollers. There were more people out in the early evening than ever walked the main street of Fremont, Ohio! He smiled to himself as he finished the espresso and dropped the cup in a waste can.

He walked up the Strand to the north a few blocks, looking at the ocean and the mountains to the North. He finally decided he was hungry enough to eat some kind of supper. He walked up a 'walk-street' to Manhattan Avenue and turned south to find The Kettle.

Amanda McCormick was seated at the bar! He smiled and walked towards her, smiling. "Might we?" he asked, gesturing to a table, as she smiled at seeing him.

She stood up. "I suppose we might! Imagine finding an Ohio boy way out here!" she exclaimed, smiling.

Grant raised his eyebrows to a waitress in black and white, gesturing towards an empty table. Moments later, Grant stood while Amanda took the chair across a small table from him.

"This is a pleasant surprise, Grant Porter, Ohio Boy," she said, still smiling a big, radiant smile.

"Uh, it's pronounced Ohia, Ohia Girl!" They both laughed. "Did you get an apartment and restaurant list, too, Amanda?" he asked.

"I did; did you go look at any rental places?" she asked, as she studied the menu.

"I did; I looked at a place called New Rameno Village and another called Tesseract Towers. They're both smallish but pretty clean; making a decision about a place out here in this megapolis is scary! I decided I was still a bit jet-lagged or overloaded with new ideas; I'll have to go back and look when I'm more alert," he allowed.

"Yes, I looked at New Rameno Village, too. There are a lot of restaurants there. I looked at some new development called River View in Diamond Bar; there's a dry canal they can see. Diamond Bar might be a bit further to commute. I don't know what I'm going to do for a car; you *have* to have a car out here! What are you going to do?" she asked, smiling that very nice smile: perfect teeth and deep, dark eyes.

"Well, I've an old Dodge pickup my Dad gave me; it's ten years old, a bit worn, now. I'm thinking about whether or not I want to try to drive it out here. It'd be good because it's paid for; I might incur some expenses getting it ready for a 2,000 mile trip, but it's paid for. That's a major consideration for a newly minted engineer; I suppose I could buy a used truck out here, actually. Might be safer than trying to drive mine out here. I dunno!" he replied, smiling.

She decided on a steak salad: chunks of tri-tip steak with blue cheese in a garden salad bowl. He decided to try shrimp in pasta with a tomato sauce; he ordered a cup of lobster bisque and a double espresso.

The still-warm dark bread came on a cutting board with a small ceramic pot of butter. "This is looking pretty good, isn't it?" he asked rhetorically, cutting two slices of the bread.

"It is! It's reassuring to have someone to talk to, Grant! Who'd have thought we'd grow up within an hour of each other and meet out *here*! What do your parents do?" Amanda asked.

"My Mother is an NP/administrator in the Fremont Promedica Memorial Hospital. She has a bunch of RNs who do the work. My Dad is a technical salesman for a line of furnaces, kilns and ovens; he travels about one week in three. What do your parents do?" Grant replied.

"My Mother's a doctor, gynecologist, at Firelands Regional Medical Center in Sandusky!" I wonder if they *know* each other?" Amanda asked, big eyes wide open. "My Dad's a city councilor; he wrestles with how to pay for muni improvements." She shook her head, sending the curls moving nicely.

"It is amazing to meet like this!" he said, not brave enough to comment on that in any depth. "So what does a ceramist intend to do with her specialty?"

She smiled. "You even pronounced it correctly; extra points for that! Where'd you learn *that* word?"

Grant smiled, considering how to answer. "Well, I read; history is... *anchored* on a time-line by the very small bits of art that survive to be found by archeologists and paleontologists: pottery, glass, bronzes and bones! Any good engineer should at least know of what our precedents were able to do! A... budding materials engineer should learn about a lot of things. Anyone interested in microelectronic circuitry should know that all circuits, chips, are made of metals, semiconductors and dielectrics. I'll bet we'll be using silica, alumina, beryllia and other metal oxides, if we go to work at NG!"

Amanda shook her head, eyes bright. "Damn! You're well-read, Grant! I don't know anyone who can put silica, beryllia and metal oxides into a sentence! Impressive! Where'd you learn 'beryllia'?" she asked.

"BeO is a common dielectric substrate used in microcircuitry because it's more thermally conductive than aluminum! You can conduct some of the chip-generated heat out of a package with BeO; Al_2O_3 and SiO_2 just don't work! Copper/tungsten is a decent conductor; beryllium metal, is pretty good, too. Copper/diamond is a *very* good conductor! A *materials* guy, person, should know this!" he said with his own smile.

"Wow; my first materials lecture over dinner near the beach in Golden California! You're... interesting, Grant Porter! My Mother would be impressed if she knew as much chemistry or metallurgy!" She chuckled and shook her head.

"What?" he asked.

"Oh... we are joining an *elite* society: those people who *know* about the physical world. In ten years we'll find it hard to talk to those who have no interest in the physical world and are more interested in their toys and malls and money!" she sighed. "I like being able to talk about material things. You said you... made some things of glass?"

"Yeah, some jewelers sinter colored glass frit, ground glass, into pendants and earrings. Glass 'jewels' are inexpensive. Did you know there's a whole group of 'jewels' made by Swarovski in glass? I also take brightly colored art-glass sheet and heat it enough to slump into a mold to make little... candy and nut dishes. It's fairly quick; you can make dishes that are pretty attractive and inexpensive. I don't think jewelers are much as engineers, but some of them are *very* creative. Swarovski is in the Austrian Tyrol; someday, I want to go there! *Someday* I want to do a lot of things!" he said with some passion.

She laughed. "So do I! That sounds like a European trip. I saw some of France, Italy, Germany and London on my 'great European trip' as a senior in high school. I didn't get down into Southern Germany," she said with a sigh.

"I have a list of places I want to see: the Tyrol is one of them!" he replied, remembering making a list of places he wanted to visit. The list had grown over the years as he read of places and cultures.

"What's on your list?" she asked, leaning forward.

He looked into her eyes and smiled. "Well, the cities of the Hanse around the Baltic in northern Europe, Andalusia in southern Spain, the islands of the Caribbean, Coastal Alaska, England maybe, France, maybe, Vancouver Island in Canada," he sighed. "Someday!" he said with a smile, holding up his empty espresso cup to her iced-tea glass.

She nodded and said: "make no *small* plans! Hey, what did you tell McPage when he asked you what you wanted to do?" she asked.

"Ah... I didn't know what to say! What does a new BS engineer know? I said I might want to go into microelectronics and pursue some metallurgy. I said I had done some composite structures work and I knew a little of hard soldering." Grant shrugged. "I don't want to be like that Gannon guy: he wants to have a title and wear expensive clothes!" Grant replied.

She laughed and then frowned. "*That* guy creeps me out! He wanted to go out to dinner tonight! I turned him down cold. That guy is a bit *weird!*" she replied.

"I told McPage I was a student pilot; I think he liked that. He said he never took the time to get a private pilot's license. I'll do that when I get my Purdue debt paid off!" and he sighed, thinking about that.

"You hurting over that?" she asked.

He shrugged. "I don't owe *that* much; I can maybe pay it off in a couple years if I'm focused. I'll have to construct a real budget once I accept a job offer. *Someday!*" he said again, smiling.

She nodded eyes opaque. "I'm... debt free. I worked as a barista in a bar in Troy; the tips are pretty good. I only worked two nights a week; I was very careful. I studied karate in high school; I decked a guy one evening when he got lippy! Surprised us both; broke his arm when he landed on a cocktail table. He never came back... which was very good!" Her eyes clouded as she remembered.

"I can see you behind a bar; engineering should be a lot more interesting!" he nodded.

"Yeah, I think so. Hey, you want desert? There's supposed to be an ice cream place down toward the beach!" Amanda asked.

"There is; it's a block or so down that way," he looked around and pointed. They paid their tabs and exited the restaurant to find that a light gray fog had blown in off the ocean.

He slipped his sports coat on and she put on a sweater she'd brought. "I didn't expect *fog!*" she said.

"Well, warm air and cold water is all it takes. Might be an advective fog: moist air forced up onto cooler land. They warn student pilots about flying uphill into cooler air!"

"'Pilot' is interesting, Grant. How's it feel to *fly* a plane?" she asked as they walked below the orangish ball of fog around the sodium vapor streetlamp.

He smiled to himself. "It's a very *different* thing, actually. You need to know a bunch of things about how the specific plane flies. You need to know what the engine and propellor *sound* like. You need to... see the plane's wings and ailerons in your mind's eye... and kinda *feel* the air moving by. You need to understand, comprehend, how the wings are working, especially as you glide down for a landing. Takeoffs are easy, landings are more complicated! I need to do a lot more before I'll be any good at it! I've soloed, but I gotta do that a lot more to really have it *learned*. You want to fly? We might do that sometime; I need to get current when I've got less debt!" he offered.

"Yeah, I might want to try that sometime...*sometime*!" she said, mimicking his use of that word, with a smile as they stood at the ice-cream 'bar' deciding what to have for dessert.

A few minutes later they walked out along the Strand with waffle cones where there were still people walking, skating and biking in the descending darkness. "Strange place!" she said, smiling as people zipped by them.

"Yeah, I was thinking that earlier when I walked down here. A *lot* of people just going by; a *lot* of young women showing a lot of skin!" he shook his head.

"Well, it's California; I think some of that is in the air, or something!" she chuckled.

"So, are you ready for an job offer, Amanda?" he asked.

"I don't know," she said. "Our resumes are so short, there's not much to discuss, is there? You ready?"

"I don't know, really. I think they have to bring us onboard and teach us some basics and see what we can do. McPage hinted at us being an 'entry class' of some kind. The interview didn't explain much about that," he replied.

"I know; I hadn't expected we'd form a 'class' of any kind. It makes sense for them to train us all together, I suppose. Are you going to ask for anything specific?" she asked.

"No; I told McPage I didn't know what to ask for. I think at some point they point us at a problem and see what we do. I studied something about aerospace, the industry. There's a bunch of people who study and design *structures*: fuselages, wings, landing gear, engines. There's another bunch of people who work on the *electronic* communications for those structures. Some of it is satellite electronics that I know NG is big in; some is just avionics. They do some of that, too. I think the microelectronic... issues are a huge field. I might ask for that. I don't know how much they need metallurgy out here; I have some background for that. Did you ask for anything specific?" he asked.

"No. I hinted that I should be able to work with substrate technology in microelectronic circuits. Some of that is just words I know; I don't know what he thought of my fumbling for words. I suppose we'll find out tomorrow morning!" she replied.

"Yeah; we all have to start somewhere, right? We better go get some sleep so we can be bright-eyed and bushy-tailed for that! Are you at the Residence Inn?" he asked.

"No, I'm over in El Segundo at the Hilton Doubletree. How's the Residence Inn?" she asked as they walked back up toward The Kettle.

"It's OK; nice fresh croissants at the free breakfast buffet thing, but I had to go to Starbucks to get a half-way decent espresso. I'm sure they, NG, get some discount for putting people up there. The clerk suggested they have some kind of long-term discount if you stay there. We may learn more about that in the morning!" he replied, very much enjoying talking to this attractive young woman.

She stopped at a small KIA rental car and he paused, not knowing if their dinner together merited anything like a kiss. He decided not to risk that and held out his hand. "I enjoyed having dinner with you, Amanda, very much! We might consider that another time, maybe."

She took his hand in her's easily. "I enjoyed it very much, Grant. Why... don't we meet for breakfast before going over to NG? You can get espresso at The Kettle or at Starbucks," she suggested.

"OK; that sounds good! Very good!" he smiled. She was engaging! "Say 7 AM, at The Kettle?" he asked.

"Yes, let's do that! I might need some caffein before whatever they're going to tell us! See you here!" And she got in her little hybrid and drove off with a wave.

He stood for a minute and then walked on up towards his car. He walked into a Von's supermarket and bought a bottle of water. Somehow, he was thirsty! He looked forward to seeing what they'd offer him. It would feel

44

good to seize some kind of job and begin doing something useful!

He drove back to the Residence Inn and spent some time looking at the rack of tourist attractions for a few minutes before going into his room. He showered and went to bed thinking about big eyes and dark brown curls. He hadn't done a lot of dating either in high school or at Purdue; he was usually too busy. He hadn't wanted the distractions of trying to manage a close relationship; it took time and money and he had too little of either! He fell asleep rethinking that conservative approach to life.

Chapter 3   Offer

Grant waited on the corner until Amanda walked down from where she'd parked on Manhattan Ave. The sky was clear blue with the sun still hidden behind the tallest dune where Sepulveda Boulevard or PCH ran north along the coast. The air was cool; the wind was gentle; the ocean had small waves of about 6 to 8 feet; a few surfers in wetsuits waited, straddling their boards; it didn't look like there would a lot of action with such small, calm waves. It felt good to see her smile as she walked up.
"Good morning! Are you ready for this?" he asked.
"I hope so!" she replied, smiling. She wore a different pair of beige slacks today and a long-sleeved, dark blue blouse under a heavier blue sweater.
They took a booth and immediately ordered breakfast and espresso. He ordered huevos rancheros, mostly to see if he liked it. He'd never seen that dish on a menu in either Ohio or Indiana! Apparently she hadn't either. She studied it when the waiter brought out her eggs, hash browns, toast and orange juice. He ordered a second double espresso before he dug into what turned out to be a very nice mixture of eggs, sausage bits, peas, cheese, celery slices, flour tortilla, some red sauce and a dab of pico de gallo!

"This is very good! I'll have *this* again, maybe soon... who knows?" he chuckled, using the napkin to be sure he didn't leave very much food on his face.

"I considered it; I never saw it before. Have you?" she asked.

"Nope! Gotta be *carefully* experimental with new foods, according to my Mother! My Dad's pretty experimental, I guess. He travels a lot more than I ever did, mostly between Chicago, Boston and North Carolina, what he calls the 'Eastern Industrial Triangle'."

Amanda sipped her espresso slowly. "I hope they'll introduce us to whoever we're going to work for. I'd like to know who I'm reporting to!" she said.

"Roger that! Hard to prepare for someone you don't know!" he agreed, suppressing some concerns about what he'd be offered and who he'd be working with.

They finished breakfast and parked in front of the R1 building just before 8 AM. He held the door for her as they went inside; she had a medium-sized purse. He carried his folio and his leather-bound iPad. They waited quietly in chairs in front of the HR offices where they could see Mrs. Miller and Amy discussing something at Mrs. Miller's desk.

Gannon came in next in another 3-piece suit; he was consistent. My-An came in behind him. Two of the other guys came in, in a rush, right at 8 AM. The third guy didn't show.

At seeing Grant looking at the door for the last member of yesterday's group, one of the two men smiled and whispered: "Danny's out! He flunked the pee-bottle! We're one down for the day!"

Grant frowned; how *dumb* did you have to be to flunk a drug screen? Why would *anyone* try that? All you had to do was stay clean for a month or so. Apparently 'Danny' couldn't do that!

Mrs. Miller walked up to them. "Amanda, please come with me; we're going to meet with Mr. McPage now." Amanda got up, glanced down at Grant, not quite smiling. Grant smiled and nodded, staying silent.

Amanda was gone for about 30 minutes. She came back smiling, making eye contact with Grant. "Grant, you're next, please," Mrs. Miller said.

Once seated in Mr. McPage's office, Mrs. Miller opened a folder and handed documents to McPage. "OK, Grant; can I get your passport number, please?" Grant handed him a piece of note paper from the Residence Inn with the number written on it. McPage input it into his big desktop comp and waited. After a moment he nodded. "OK, Mrs. Miller; you might go chase this through the system while I explain what we're going to offer Grant."

Their computers were linked in some kind of *intranet* that let them see the same documents. Mrs. Miller left the office through a side door and went to her desk.

"We liked your grades, Grant; a solid B+ in the old lexicon! We liked that you worked part time in a jewelry shop and in the machine shop at Purdue. We like that you see yourself as someone who wants to *build* things! So do we! Incidentally, although she didn't put it quite that way, so does Amanda McCormick! You might work with her on some of what we see ahead of us, if you both accept our offers," McPage said with a smile. Could McPage see his interest in Amanda? Or was it that most males would be interested in such an attractive young woman?

"Our offer is $62,000 the first year; that's $5,166 per month or about $21.57 per hour, if that helps. We consider you on probation for the first 3 months; after 90 days, we'll review your contributions and start ranking you with your peers. Your initial security clearance should be complete at that time. We sometimes bump your pay if we like what we see. Automatic adjustments run in the 2 percent range. So, you might see $63,240 per year at that time. An annual review and reevaluation at one year might be about 2 or 3 percent; so after one year's employment, you might expect to see about $65,137. If you do something well, you'd expect and receive more. Some hard-starters might see something closer to $70,000 at the end of their first year." McPage explained and paused.

Grant did some quick arithmetic in his head. The offer was better than the investment casting firm's offer; it wasn't an extravagant offer but it was *interesting*.

"Included in the salary are full medical, dental, vision expenses. You'll be scheduled for a medical exam, what we call an 'incoming appraisal'. We want to know, and so do you, of *any* issues before they become significant. We have a small bonus program to help remind you of the critical necessity of controlling your weight. NG is a no-smoking facility; you don't smoke and we considered that."

"We will pay to have your clothes, books and your car shipped out here, however you want to do that. That includes your expenses in getting out here, should you want to *drive* your goods out here yourself. Amy said she sent you to look at singles housing; did you?" McPage asked.

"I did; I found a couple places out off the 91 that might work. I... might want to drive myself out here; I inherited my Dad's old pickup. I might make that do for a while," Grant replied, worried about commuting 2 hours a day in the old truck for an extended period of time.

"We'd pay fuel, oil, tolls, mileage and both food and motels on the relocation move. Field hires are more expensive because they have accumulated more stuff and often have families! We'll put you up at the Residence Inn for 2 weeks while you locate a permanent residence. Mrs. Miller hasn't turned up anything the Feds are interested in on your

50

passport. That's very good!" They had some kind of internal intranet for their computers that was linked to outside services!

"You noticed that Dan Burney, one of our two industrial engineering candidates, is *not* with us today. He failed his drug screen; that's unfortunate, but not that unusual, especially in certain elements of our society. We do not tolerate illegal drugs, marijuana or alcohol during our standard work day. We do not hire illegal residents. One of our first entry-level class sessions will be to introduce both Federal security concerns and issues and what corporate proprietary information is and is not. You know how to research *public* information channels; we have our own proprietary information system linking our facilities within the States. You will be encouraged to use that system to your advantage; there are occasions when you might be invited to instruct video classes in some technologies using that channel. I think you could do a session on lost-wax investment casting with no preparation, Grant!" McPage was emphatic on that. He was surprised the man liked that; he might well do such a thing.

"We pay full tuition, books and fees for technical instruction at selected universities such as USC, Cal Poly and UCLA; we allow you to flex your schedule to attend regularly scheduled technical classes. We have some hooks at USC and UCLA; they are both within commuting distance from the South Bay. We encourage you to go for a masters; we will 'adjust' your salary once you obtain a masters or a doctorate. We pay a bonus for patents granted. I want you to consider continuing to study; it'll make you a more capable engineer! Questions?" McPage asked.

Grant took a deep breathe. "How long do I have to consider your offer... which is consistent with another offer I've received back home. I'm not sure I want to live there; but I could do laundry at home there!" Grant said, making a feeble joke.

"We'd like to sign acceptance papers today or tomorrow if we can. We do schedule a few weekly and monthly training classes for new hires. We're always adjusting the timing of those classes, but it's easier once we have you on board. I expect you might like to meet with McCormick and Nguyen and maybe Gannon to hash over what you've just learned. You might do that here; there's a small group of high school seniors, a dozen, touring our Manhattan Beach campus this afternoon. We don't take visitors through D1, our chip-making facility, but we do take them through our Hybrid Microelectronics Lab in M5. You and McCormick might find that most interesting. I assume you know some of the *vocabulary* of microelectronics?" McPage asked.

"Yes, I do; one has to, to be an engineer today!" Grant said with a small smile.

"That's more true than you may know, Grant! Take time for lunch; meet Amy in the smaller cafeteria over in M5, across the street at 1 PM. She'll combine you with the high school tour. If you've never seen modern chip assembly, you *need* to see this!" McPage declared. And McPage stood up to shake hands; the offer was complete. It was us to him, now!

Grant stood up. "Thank you, I'm considering your offer!"

"Good; you're going to learn to 'drink from a firehose' here, Grant. Enjoy it!" McPage replied.

Grant went outside the office and found Amanda waiting for him in the HR lobby, smiling. "Are you smiling?" she asked.

"Yeah, I guess. Are you up for espresso? I think I could use a Starbucks about now!" he declared.

"Roger that!" Amanda said. "I'll drive. You want to make Amy's tour, don't you?" she asked.

"I guess I do; he mentioned you to me! Said we might end up... collaborating, I think," Grant said with a smile.

She nodded as they went out to her rental car. "Neither McPage nor Miller explained how formal their training classes are. He mentioned we might collaborate, being a 'materials guy' and a 'ceramics girl'; that might be interesting. You rattled off some substrate materials at dinner. You ever work with any of those?" she asked as they got into her car.

53

"I haven't; they're mostly words in a table in a textbook. I expect we might run into them if we take these jobs. My offer was a bit better than the casting house's offer back home. Part of that might be cost of living in Southern California; part of it might be to hire people they think might actually do something. Do you know if Gannon got an offer?"

"I don't; he went last after My-An and the other two guys. Were you shocked that Burney failed his drug screen? *I* was. That was really *stupid*! *My* offer was pretty good, but I'm not going to compare it to yours, woman and all, that I am!" She was very definitely a woman!

They sat at a table in the Starbucks across from The Kettle, a few minutes later. "Well, I like it; I did some quick calculations; if I spend $900 a month for a small apartment, I ought to be able to save a thousand a month; that's the right thing to do, if I can manage it. I'm thinking we might eat out most nights, right? I'm a... lazy, indifferent cook," she confessed.

"You used a very interesting word just then, Ms. McCormick!" he began with a small smile. "'*We*'. We *might*, actually! I'm going to tell people *you* asked me out," as she colored and then smiled. He smiled a much bigger smile as she shook her head.

"OK, I want to do dinner together tonight, at a place with some freight, maybe a

Chart House or a McCormick and Schmick's. You all right with that?" she asked, smiling.

"Absolutely! 50-50, but that's a *good* idea!*Maybe* I should accept this offer because NG has so many opportunities we might take advantage of. Did you like the offer to pay tuition, books and fees for technical courses on flexed time? That's very generous!" he exclaimed, beginning to think this might be a good way to start his career as an engineer.

"It is and I did! That would be a very good longer range plan, actually; start on a masters as soon as possible and keep plugging away, course by course over time! That's a pretty easy way to guarantee a raise at some point, I guess," she said, looking off in the distance for a moment.

He did some wool-gathering of his own for a few moments, trying to look into the future.

"What?" she asked seeing his face go blank. "Us! Did we just become an 'us"? he asked, catching one of her hands on the small table.

She tipped her head to the side as she held his hand. "That's scary... maybe so." she allowed with a small smile.

※

They talked about nothing for a while before grabbing a quick lunch at the McD's on the corner of Aviation and Manhattan Beach Boulevard.

They went to the Guard Shack at M5 and asked for Amy Borden who might be in the M5 Cafeteria. Several minutes later, Amy waved them past the security guard on duty. Their temporary badges wouldn't let them into the building without her permanent HR badge.

"We're going to gown-up, put on spittle masks and booties. We're going to keep our hands off everything, mostly tables, to prevent ESD damage to microcircuitry. We're going to stand back from the tables and just *watch*. Most of the equipment is labelled as to function; I'll answer questions, if I can," Amy explained.

"The Hybrid Assembly Lab is where semiconductor chips from our D1 foundry are assembled into packages made in M3 and M4. These packages are custom designed by our EEs. The packages are built in-house; they are all proprietary or classified secret. We won't talk about them in any detail until you're cleared for that discussion. The display screens above some of the assembly equipment will let you see some details of what they do and how they operate. I know only a very little about what we're going to see here."

The 15 of them went into a 'change or dressing room' where they opened white Tyvek ESD 'gowns' and put them on. They put on ESD foot-straps to conduct stray charges to the conductive tile flooring. They donned white paper spittle-masks. "One of your first class sessions will be to teach you about electrostatic discharge physics, 'ESD'. The friction between the clothes you're wearing and your skin, generates charges, high voltage charges, on your skin and clothes. If you touch a package, that charge can zap through the circuit, frying the chip. All the operators, techs, and engineers wear both wrist and shoe straps, and the work tables and chairs are grounded to prevent ESD. It's complicated, but we know how to do this. The chips are very small and have correspondingly small circuit conductors: 'wires'. 30 years ago, all the circuits were bigger and more robust; today they are orders of magnitude smaller and more sensitive to ESD. Don't *touch* anything in the Lab, please. Do you understand what I said?" Amy made eye contact with every one in her tour group and then led them inside the Lab. "Just watch, don't touch. We'll answer questions when we're done!"

Amy got Amanda and Grant at one end of her tour. Grant recognized one of the bonding machines. "Do you know what we're seeing?" Amanda asked him.

"Yeah, sorta! These are 'die-bonding machines'; they stick a semiconductor chip down on a substrate. The white... 'floor' of those packages are substrates: either alumina or beryllia. The die-bonders *position* the hybrid components on substrates: semiconductor chips, resistors, capacitors, inductors. The next machines will be 'wire-bonders'; they will... *wire* the hybrids into a circuit. That's an *aluminum* wire bonder; see the big spool of wire, its *silver* colored? That's a *gold* wire bonder: its wire is drawn gold. I'll explain later! This is the Silicon Age; the bluish chips are... silicon or maybe gallium arsenide. Each of those chips is like what's in our cell phones! Damn! I've seen pictures and even a video; this is very fundamental to our modern... culture! I think I'm going to accept the offer! I want to learn *this*!" Grant exclaimed through his mask to Amanda, excited to see something us close and personal, instead of in a textbook!

Amy walked them along long rows of tables where ESD-garbed operators worked silently, assembling an array of different packages on the highly automated equipment. The wire bonders worked like tiny sewing machines, bonding wire from package to chip and to the next chip in a blur of very small, precise movements. The gleaming wires could be seen like spiderwebs, smaller than human hair, in the intense lighting arrayed around the packages.

Two techs and a tall, thin engineer, Grant guessed, stood at a big comp display, pointing to details in the image in front of them. He glanced around the room and located a side room with a sign that said: 'Epoxy Storage'. Another sign read: 'Laser Welding' Another sign read: 'Hard Soldering'. There was a *lot* of technology being performed in front of him; he had a lot to learn! NG might be a good place for that!

Amy led her tour back to the 'Change Lounge' and they removed the gowns, masks and booties. She led them back to the small cafeteria the assembly operators used. "What do you think, Amanda, Grant? Amy asked.

"I've got a lot to learn; I've seen photos of some of those machines; you use both aluminum and gold wire; there's a lot of metallurgy involved in how those machines work. Very interesting!" Grant replied, feeling more and more comfortable about accepting the offer.

One of the high schoolers asked: "so those technicians put a semiconductor chip down and the wire-bonders connect it to other chips?"

"Exactly!" Grant said, before Amy could answer. "Each chip is a circuit; the wires connect chip to chip and chip to the other components. The white rectangle in the floor of the... Amy, am I OK?" Grant asked.

"You're doing fine! You're the engineer!" Amy replied smiling.

"What's a substrate?" the girl asked.

"Well, it's from the Latin meaning underlying floor or base. In the electronics sense, it's a... dielectric foundation upon which the microcircuitry is placed. The substrates we saw were either alumina, Al_2O_3, or beryllia, BeO. Alumina is less expensive but it doesn't let chip-generated heat get out. Beryllia, BeO, is both a dielectric or *insulator, and* a thermal conductor; beryllia has 200 times better thermal conductivity than alumina! BeO is toxic, more so to some people than others. So, we need to not abrade it where we might breathe the dust. The microcircuitry needs to be *separated* from the electrically conductive metal of the package but allow some thermal transfer to get heat out. It's a complicated interplay of requirements, especially at the high frequencies one assumes these circuits will operate at: maybe 20 to 40 Gigahertz," he guessed.

"Why don't those rows of wires in the package... ground out the electricity?" the girl asked.

Grant pointed to a big photo-poster on the wall of the cafeteria, one of several: it showed a magnified stereophotograph of a package. "See the circles around the wires here? Those are *glass* tubes called *feedthroughs*; each wire through the sidewall of the metal package is insulated from the metal wall with a very small tubular piece of borosilicate glass. Glass is a very strong, very cheap, electrical insulator. There's a lot of physics and some chemistry involved in feedthroughs! Google it!" he suggested as the girl smiled.

"You sound like an engineer! You work here?" she asked.

"Not yet! Maybe soon! I'm a new Materials Engineer; I study an amazing array of materials and how to use them! Your cell phone has a single, big microcircuit in it. If we opened up your phone, we'd see something like that image!" he said, pointing.

"You can have my job, Grant!" Amy declared, smiling.

Grant smiled. "I can't; I need my *own*! Thank you for getting us in here! If we want to *accept* our offers, we do what?" he asked Amy.

"You go back to R1 and sign some stuff with Mrs. Miller!" Amy replied with a smile. Grant looked at Amanda, who had taken his hand in hers. That was interesting!

The two of them walked back outside through a door that Amy led her charges through. "I'm walking them back to RI, that way. You probably want to drive!" Amy said, with a wave.

"You've seen some of that equipment, assembly equipment, maybe, in text books?" Amanda asked.

"I have; it was very good to see the actual equipment up close!" Grant replied.

"You know of glass-to-metal feedthroughs; I thought that was a fascinating development: Westinghouse's Kovar alloy and Corning's borosilicate glass!" Amanda said, looking at him in the afternoon sun. "You want to go sign up?" she asked, studying him.

"What do you think, Amanda?" he asked.

She nodded for a minute. "I think this is a good place to start earning our salaries, Grant. Let's do it!" she squeezed his hand with hers.

"*Damn*, I suppose we should. We need to figure out how to get home, pack and come back and how long that's going to take, don't we? 2,200 miles is going to be... 4 days at 550 miles a day. Better say 5 days, in the case of my old truck, which may malf on me coming west! Would you just fly with your stuff as additional baggage?" he asked.

"I don't know; I haven't had time to figure that! I might get some boxes and have it shipped out,
Fed-X or UPS, maybe. I have no idea what I'll use for a car out here! I suppose I could rent one for a while; hey, maybe you could help me find a beater to buy?" she suggested, looking into his eyes.

"Yeah, maybe... sounds like we have some planning to do, huh?" he replied, slowly, smiling at her.

"Let's go talk to Mrs. Miller; she's seen this done before. I need to make a time-line!" Amanda said, smiling.

"Roger that!" Grant said, smiling into very big, perfect eyes.

Chapter 4   New Jobs

Amanda led Grant into the R1 lobby and off to the left toward HR. Mrs. Miller stood up and beckoned them into her office. "So, what have you decided?" she began.

"We... I want to accept and plan how to report for work. I'd like to know who I'm going to work for and meet them, if we can arrange that. I think Grant would like to do the same," Amanda replied.

Grant nodded, silent for the moment. "All right, Amanda, we're going to assign you to our Materials and Processes group. They're located in R6, just west of M5 where you got a quick look at Hybrid Assembly; those 2 buildings join together. The group's manager is a man named Dr. Howard Morgen; I think he's a graduate metallurgist; his secretary is Alice Stapleton. She's the person who will get you a desk, a chair, a computer and get you hooked up to a printer. You want to do this, Grant?" Mrs. Miller asked.

"I do. Who are you going to assign me to, please?" he asked.

"Robert has arranged for you to report to our Manufacturing Engineering unit under a woman named Paula Hopkins; her secretary is Sali Tompkins. Manufacturing Engineering is a newer unit; one of their tasks is to introduce new technology to the various production lines. I think Robert would like to see you! He's beckoning us to come over there!" Mrs. Miller gestured. They walked around to McPage's office.

"So, are going to join us?" Mr. McPage asked.

The two of them nodded. "Very good; I think you'll enjoy working here! Amy said you explained about hybrid assembly to one of her tour group's seniors! I'm going to work Grant's induction; Leona will do yours, Amanda. Let's get started!" he said, smiling. Mrs. Miller handed a packet of papers to McPage and gestured to Amanda to follow her back to her own desk.

The induction procedure was involved; it took almost 40 minutes, with McPage explaining each form to Grant as they went. Grant had to sign an acceptance form with a lot of legalese. There was a security agreement that allowed NG to do a formal background check on him. There was a series of medical, dental and vision insurance forms; there was a patent release and a form that swore him to hold certain privileged intellectual property 'in trust'. The forms all used his parent's address as his legal residence.

"We update these forms once you establish permanent residency out here. Part of that residency will be to get a California driver's license and insurance coverage. You'll want to open a bank account out here; we'd like to directly deposit your salary into that new account. It's quicker and safer," he explained.

"I think we can introduce you to Mrs. Paula Hopkins in about 10 minutes; she's over in a meeting in M3. We'll head over there. I'll introduce you to Sali Tompkins, Paula Hopkin's secretary. You'll be upstairs above and in front of the main Mechanical Manufacturing shop in M3. Let's walk; we'll come back here afterwards so you can catch up to Amanda."

"A word about Manufacturing Engineering; this newer unit is supposed to function as a *bridge* between engineering and manufacturing. Those words carry some freight, of course. Manufacturing is supposed to *build* things. Manufacturing Engineering is supposed to introduce *new* materials and improved processes to Manufacturing to improve how they build. Sometimes Manufacturing Engineering kind of hold's Manufacturing's hand while they mutually learn how to do things, especially *new* things," McPage explained.

"Different corporations do things in different ways; Amanda's going to start in what we call Materials and Processes or 'M & P.' They are... *developmental* people who assist Manufacturing; some times they help put out fires. Sometimes they go on the line to augment new materials, sometimes with Manufacturing Engineering. I think Amanda's going to be doing some comparisons of substrate capabilities initially. We are trying to reevaluate some fundamental capabilities in the building of ceramic substrates in house:

alumina and beryllia, for the bulk of what we do. We used to have a thick-film unit that did screen-printing of metal conductors, inductors and resistors on alumina; they did the firing of glass frits and pastes over in R6. We had an issue with beryllia poisoning and some management issues and we shut the unit down. We're going to reestablish most of that capability with an outside specialty group. I believe Amanda's going to be involved with that," McPage explained.

"Manufacturing Engineering may be involved as well, but I know Paula may hand you a new bonding material... I think it's called 'SuperCon'; it's a diamond-filled epoxy film. She's been poking me to find someone she can hand it to. You may get to introduce this new epoxy to our assembly line, Grant!"

Grant nodded as McPage badged them into M3. "We'll get you a permanent badge when you report in; the badge will allow you to enter certain buildings you need to work in. We'll add buildings to your badge as you need to enter to attend meetings or go out on the manufacturing floor. I'm guessing you'll spend some time in both M3 and M5 and maybe R6. We have your cell phone number and may call you if anything comes up on the papers you just signed."

A medium tall, dark haired woman in dark blue slacks and a white shirt, approached them. "Paula Hopkins, this is Grant Porter, a materials engineer from Purdue who just accepted our offer. You may want him to work on SuperCon!" McPage said with a smile.

The two of them shook hands, studying each other for a moment. "It took you long enough, Robert! I need to discuss our unit's composition and tasks, Grant. You're going to see if this new and expensive epoxy bonding film is good enough to introduce to our high frequency hybrids. You know anything of epoxies?" Paula asked.

"I know a little; I've made some experimental aircraft skins: graphite/epoxy, glass/epoxy... hand lay-up. So this SuperCon is a diamond powder dispersed into a precast epoxy film?" Grant asked, guessing.

"Very good; it is," she smiled. "Diamond dust is a new commodity coming out of China! They've learned how to make it marginally affordable. *If* it'll help us conduct heat out of the base of hybrid packages more efficiently, we want to learn how to use it. I need someone to help figure out if it's a *real* improvement or just hype. Does that sound interesting?" she asked with a smile.

"Yes, Ma'am, it does! This is a 'die-bond' or substrate bonding film?" he asked, trying to remember the correct wording.

"It is; about .002-inches or .003-inches thick... premixed and frozen, stored below minus 40-degrees F. One thaws it at room temperature and razor-cuts it to shape. Its vacuum-bagged to press the substrate against the adhesive film and heated to cure in a single cycle. We've done some preliminary testing but nothing definitive; we want you to define it for us. When can you start?" Paula asked with a quick smile.

"Uh, in something like a week, Ma'am," Grant replied, hoping that was acceptable. She nodded.

"He's moving out from Ohio; can I get back to Sali on that? I'll give her his cell and vice-versa. He can determine the start date between us from the road, if need be!" McPage looked at Grant, who nodded agreement.

"Sali Tompkins, my secretary, is upstairs in M3-210; go introduce yourself. She'll get you squared away. Let her get you a new badge so you can get into our little office bay up there. I hold a Monday morning unit meeting at 8 AM weekly; we go over our unit's list of 'to-dos' and problems as a group. We've got a nice list of issues to be worked! Good to have you come aboard!" she said and was off out the door to another meeting.

"Go upstairs, meet Sali; I have to go back to my office. Don't worry about the badge for now!" McPage said and he was gone.

Grant went upstairs and met Sali, a youngish blond with long hair in a ponytail; she showed him a three-person office and pointed to a desk. "You can have that one; I'll order a desktop computer from IT for you. Did Roger explain about personal comp's?" Grant shook his head, no.

"I'll get you one of our iPad's; it's what you're allowed to use here. You have to lose that one of yours; we don't allow personal comps in any of our buildings! They need to be NG-dedicated, proprietary; they stay here. You lock them up at night so no one else has any access to what you put in them!"

"OK; that makes sense, I suppose," he replied.

Grant went back to R4 and waited outside Mrs. Miller's office for Amanda to show up. Might she agree to *drive* out here with him? That would certainly keep him more awake and allow him plenty of time to get to know her. He decided he wanted to get to know *her*, very much! She had shown some interest.

Mrs. Miller got Amy to get him a new temporary badge that allowed him into both M3 and M5. "You hang on to this, Grant! It'll get you into your office and into the Hybrid Assembly area where you'll be working the new material. Your contact in M5 will be one of the techs named Jerry James; he's one of the Hybrid Assembly group's epoxy film specialists. He's done a few tests on the new material, but you'll have to work out how you want to test the material. Jerry works for one of their engineers, Larry Elgin; Larry E. is a chemical engineer from U of Illinois, Urbana-Champaign, I think," Mrs. Miller explained.

Grant thought about beginning work here; he would have a lot of things to learn, but he felt good being able to understand some of it! Amanda walked in after another hour. "You look tired, Woman!" Grant began, with a smile.

"They walked me all over M5 and R6! I'll *never* remember all the people I met! I'm sharing an office with an EE from Worcester Poly; she's black and pretty sharp! Aren't you tired?" Amana complained.

"I'm fine; I met my new boss," he said smiling. "She's going to have me work on a diamond-filled epoxy film; I got an office but I didn't meet any of my office mates yet; I'm in an office for 3 over in M3."

"I need to sit down and drink something!" Amanda said, sitting down in one of the chairs with a flourish, leaning back with a loud sigh.

"We could walk over to the Cafeteria for iced tea or cola or just water?" Grant asked, sitting down beside her.

"OK, let's do that. Amy, do we have to do anything more here, today?" Amanda asked.

"No. You're good to go for now. You need to keep me informed about when you're going to come and report for work. Day's other than Mondays are good for that; most section heads have weekly meetings on Mondays, but that varies some. Tentatively, you're both going home to Ohio tomorrow and you'll be back out here next week, right?" Amy asked.

They nodded in unison; they hadn't done any planning or tried to get flights out yet. That would have to be the next step. They walked over to the Cafeteria where they sat down and drank diet cola. Amanda had her tablet out and open. "I can go back to Cleveland through St. Louis on United at 9 tomorrow morning; that gets me home 3 or 4-ish. How are you getting home, Grant?" she asked.

Grant opened his iPad and went on Safari. After several minutes he said: "it looks like I can go back to Columbus on American... that's a direct flight, but it's a drive up from Columbus." They went back to R4 and the HR offices.

"Hey, Amy, can I charge a shuttle drive from Columbus to Fremont, Ohio?" Grant asked.

"You can; you both can: the interview trip is from your permanent residence to here and home again. I'll show you how to do the paperwork to get you reimbursed when you get back here. If either of you drive back out here, keep good notes: miles travelled, tolls paid, gas receipts and motel bills. We pay a daily expense for meals on the road; just keep the receipts, please," Amy replied.

It was almost 4:15, too early for dinner. They decided to go down to the Creamery on Manhattan Beach Boulevard for ice cream cones. They found a bench on the Strand and sat down in the shade of a palm tree to enjoy the treat and relax.

"This is almost unreal! I've got a job in Southern California!" Amanda said, smiling. "I guess we come back and start work, but we'll have to rent places as soon as we can. I suppose we look in the evenings and weekends?" she suggested.

"There's that 'we' word again!" Grant said, smiling.

"I'm becoming pretty comfortable with the word, Grant! Aren't you?" she asked with a chuckle.

"Yeah, I *am*; but it's still pretty new, isn't it?" he asked, enjoying talking to her very much.

"Yes, new word, new place, new job... new questions. We might continue the discussion over a nice air-conditioned table at an early dinner, one supposes." She looked at him out of the corner of her eyes, half smiling.

"You have enough energy to go to an early dinner?" he asked, teasing.

"I think so; we don't have to dress up out here; 'business casual' is OK. M & S or Chart House?" she asked.

"You pick, I'll drive; I've never been to either. They're both expensive by my lights!" he replied.

"I got directions from the lobby this morning. The closest Chart House is a couple miles south in Hermosa Beach at a yacht harbor. I've got it on my iPhone. We can leave my car here; I don't want to stay up very late. I'm going to book a seat on that flight!" and she was busy on her phone.

A little later Grant pulled into the Chart House parking lot and they walked into the cathedral ceilinged building. Grant admired a sculpted mahogany shark mounted on the wall. The architecture was vaguely Polynesian. They were early enough they got a small table by the glass wall overlooking the marina. Grant enjoyed looking at smallish sailboats in the very small harbor behind a seawall; not a good place to be trapped in a storm, maybe.

They studied the menu with ice water and a little loaf of dark bread. "Do you drink?" Amanda asked quietly, leaning forward towards him.

"I don't, much; it's expensive. I see they've got Sangria in a pitcher. I might drink some of that if you would. We shouldn't drink much, Amanda," he suggested.

"I agree with that, but we should celebrate our new jobs! I'll drink Sangria with you. We might have espresso before we drive back to NG. OK?"

They ordered Sangria and then she wondered if the lobster might be relatively 'local'. She wouldn't order one in Sandusky Ohio, but this was Southern California. He chuckled and agreed. They ordered a mixed green salad and more bread.

"The lobster is going to be langouste, not Homardus, especially if it came up from

Baja. I've only tasted the southern lobster once in Cleveland; I'm interested in your take on it, of course," he said.

"OK, explain, please?" she asked.

The waitress brought the Sangria; it was good: dark red with lemon, lime, orange, cherries and grapes and both sweet and tart! "This is good, but we need to not finish it!" Grant declared.

"Here's to us!" They touched their glasses together. "So tell me about lobsters!" Amanda asked.

"The New England or temperate lobster is a bottom dweller; the *langouste* is a tropical, free-swimmer and can be found in very shallow water. It has no big claws; all the meat is in the tail. The flesh of both species is white and usually served with butter," he explained. They sipped Sangria and ate some of the dark, warm bread.

Their salad bowl came out next and they served themselves from it into smaller bowls. "You excited to be employed?" Amanda asked.

"Yes, I am! You too, I'm sure!" he asked, studying her eyes across the table. He was relieved to have accepted a job offer; it was a beginning!

She nodded. "You got an assignment already; my boss just had his secretary run me around meeting people. My office is up on the 4th floor: I've got a view outside facing north...

Building D1, their semiconductor foundry," she explained.

"Mine is an interior office: no windows, no view!" he replied, with a frown. "It'll do! I don't know how much I'll be at my desk, of course."

"So what did your boss... hey, you OK working for a woman?" she asked.

"Of course I am! She sounded focused and quick, telling me what she wanted me to start on within seconds of meeting me! I think she's good; I guess I'll find out. Her secretary is younger but she seemed OK; she got me a desk and chair. Old wooden desk; the chair is wire mesh, so it's newer. She'll get me a computer when I get back. We have to use NG tablets, iPads, to take notes and carry documents around, I guess. They stay at work and are proprietary from the get-go." She nodded to that.

The lobster came out on a big platter; there were two smaller lobsters on each plate, curved in a big ellipse. "Langouste?" Amanda asked.

Grant nodded. "You've had lobster before?" he asked.

"Not this kind! I use the small fork to pull it out of the shell?" she asked.

He nodded, doing it. The lobster was white, firm and almost sweet. "This is *good*, Amanda! D'you like it?" he asked.

She nodded, tasting it. "Its good; is it maybe a little less dense than the other kind?" she asked.

"I don't know; that might just be individual variation or how fresh they are. Try the lemon juice; sometimes that gives it a nice finish." He smiled across the table as the sun started to dip behind a cloud bank far out at sea to the West. "This is very nice, Amanda. It was very good to meet you, way out here in the West! I hope you're enjoying it, like I am!" Grant asked.

"I am; I'm wondering what to say to my parents. You going to mention me?" she asked with a shy smile.

"Yes, of course... tall, slim... big, dark eyes and brown, dark brown, maybe even auburn, curly hair. Speaks well, knows enough chemistry to understand oxide, dioxide and trioxide, Probably pretty sharp... yeah, I'm thinking of what to tell them," he said with a grin to see how she'd react to that.

She laughed. "I hadn't gotten *that* far! I guess that's... pretty good, Grant! You want to work on that some more?" she invited. "You're very good with words; I enjoyed you explaining to that high school girl about substrates and feedthroughs. I don't think there's any class in most high schools that teaches about microelectronic packages," she said as she worked more lobster tail free of the shell.

After a moment, she added: "I uh, like you clean-shaven; too many guys are just too lazy to shave. It doesn't take any particular skill or intelligence to grow hair. You always shave?" she asked.

"Yeah, its part of getting up!" he replied. He thought about flying back home. "I'll have to get my Dad to help me get the truck ready for the trip out here. I haven't had it serviced in... some months, I guess. I don't have a lot of stuff to bring. I put all my books from the dorm in boxes, my clothes in a second suitcase and left them on the floor of my closet at home. I better sort through them before repacking them. The truck's got a weathered tonneau cover; that'll keep most of the rain and dust off them. How much stuff will you bring out?"

"I'm thinking about that. I better look into more slacks and shirts; most of the women I saw were wearing them. I saw some pantsuits; I saw some jeans and t-shirts! I wouldn't wear those to work! Hey, did those apartments you looked at have washing machines and driers?" she asked.

"Uh, they... had a laundry down in the basement, I think. I think maybe some of the larger apartments had those stacked, dual appliance things in the kitchen. I should have taken more time looking. In retrospect, both Tesseract Towers and New Rameno Village were pretty nice. I'll have to go back and look again when I'm not so distracted. I was tired from the flight out!" He paused.

"Maybe we should go together; you'll see things I didn't" he said, thinking maybe they should rent rooms near each other and commute together.

"Would you consider renting *near* me? We could commute together... save the cost of a car. I've *got* wheels, I just need to get the truck serviced and maybe buy a couple of tires," he suggested.

She considered the suggestion for a moment. "Let me think about it. This is pretty rich with all the butter; I like it! Good selection, Grant! Don't you need to reserve a seat?" she asked.

"Yeah, I do; I'll do that as soon as I get back to the motel! You're all set?"

"Yeah, I did it while I was waiting for my boss to get off the phone; I need to see if the airline accepted my voucher. You want dessert?" she asked.

"I was thinking of The Creamery. We should be able to find a parking space on Manhattan Beach Boulevard by the time we get back up there," Grant suggested.

"Let's *do* it!" she replied with a smile. He thought maybe she was processing something in her mind as they drove back up to Manhattan Beach. They both had a *lot* to process just now!

Chapter 5   Going Home

He idly touched his iPhone in its holster on his belt as the plane's engines reduced thrust and began it's decent into Columbus airspace. He considered calling her; they had exchanged cell phone numbers. He might have tried to switch flights and go back into Cleveland with her, but that was always a paperwork hassle with a penalty for changing flights; maybe it was better to have time to 'process' his thoughts by himself. She was disconcerting!

She was also stunning! She certainly knew she was, in a very quiet way. That was a lot of baggage to have to carry around all day! Heads turned whenever she entered a Starbucks or restaurant. Beautiful women made their own space! He had to focus on *his* new life, new job! He mused as he waited for the plane to empty out ahead of him.

Conductive epoxy: that was contradictive...almost an *omen* for his newly chosen career! Epoxy wasn't conductive, so you had to 'fill' it with metal particles to force any kind of conductivity! The common approach had always been either silver or copper. Copper *always* interacted with moisture, especially in contact with aluminum. Silver was better, but even silver could oxidize over time. Gold was sometimes used because it didn't oxidize, but it was expensive.

Diamond particles were going to be a *lot* more expensive as well as more conductive.

40-odd minutes later, he saw his Dad's new Dodge pickup at the Departures curb. He swung his roller bag into the truck-bed carefully and opened the passenger door and stepped in, beginning to relax.

"So how did it go, Son?" his Dad asked, smiling as Grant sat back and buckled up.

"It went *well*...maybe even very well; NG offered me about 11 percent more than Mercury Casting & Development. Part of that might be West Coast wages; part might be the additional expenses of living in L. A. Their benefits package is a bit more complete than Mercury's; they pay for your post-graduate studies: books, fees and tuition *and* they allow you to flex your work schedule so you can take regular classes. *That* might be a very good deal! I accepted the offer; I met my new boss and she gave me an assignment in the first 5 minutes! I think she's pretty technical!" Grant said, smiling.

"Hey, that's *good*! You gotta start somewhere!" his Dad declared as he pulled into traffic leaving the terminal.

"Yeah, I do!" Grant agreed, smiling. He wondered how to tell his Dad about Amanda? Maybe it could wait.

"When are you going to start?" his Dad asked.

"Well, I'm thinking I'll need to drive my truck out there...maybe next week, if I can get it to go that far! We might do an oil change, filter and plugs; I know the right front tire is bald. I'll have to buy at least one tire!" he replied.

"We might use Dean's hoist... inspect the brake rotors and pads. How old are those tires?" his Dad asked.

"Ah, I don't know; there's some tread on the rear tires; the front ones are... worn! I've got about $900 in my account. I might better *invest* in new tires as insurance. I'll have to lease an apartment out there; that's going to cost about $2,400 up front; the tires will be at least $500 or $600. Can we swing that? I can pay that back in a month or so." He studied his Father's profile as he drove north on Ohio 23 out of Columbus.

"I'm sure we can! How many miles are on the truck, now, Grant?" his Dad asked.

"It's 90-some thousand miles; I'm not sure. I'll probably be about an hour's commute east of NG in Redondo Beach, maybe in Orange or Yorba Linda. I looked at a couple of apartments out there, but I had too much on my mind to really focus. I think one of those places will work fine," he said.

"You might want to just buy a newer used truck, Grant! It'd cost a bit more, but have a lot fewer miles on it!" his Dad suggested.

"I thought about that; I think it'll be OK; it's not a diesel, but I still ought to be able to get 200,000 miles, shouldn't I?" he asked.

"Yeah... maybe. It hasn't been crashed or rolled; does it still track straight?" his Dad asked.

"Yeah; I think the alignment is fine; we might drive it and check that. It's going to be about 2,200 miles driving out there," Grant said, shaking his head.

"Four or five days, for sure. You won't have a lot of stuff to move, will you?" his Dad asked.

"Nope; most of it's books, still in boxes up in my room from West Lafayette. I gotta sort some clothes; from what I saw, the dress is pretty casual. Some of the engineers wear suits and ties. I think pinpoint cotton shirts and casual slacks will cover it... maybe one of my two sports coats for meetings. In the labs we're wearing ESD garb, booties and caps anyway!"

At dinner, his Mother wanted a full recount of the interview trip. He started to comply when his cell phone rang! It was Amanda! He got up and went into the kitchen to answer. "Hi there! You good, Amanda?" he asked, smiling at hearing her voice.

"Yes, I am! I wanted to be sure *you* got home OK. I know its suppertime but I wanted to call before it got too late," she said. "I missed the chance to talk with you!" she said.

"Roger that! Me too! Hey, let me call you later; what's too late?" he asked.

"Oh, maybe before 10; I'm tired. I started sorting clothes but I'm too tired to concentrate," she replied.

"I'll call before 10; I'm tired too… too tired to start on the truck tonight! Suddenly we're *busy*!" he said with a chuckle. It was good to hear her voice!

"Yeah...I guess we are. The lobster was very good; we should do that again sometime!" she replied.

"We should! I'll call you!" he said as she said goodbye!

He went back to his chair at the dining room table, smiling to himself. His Mother shook her head and made a gesture for him to explain, pulling with her finger.

"What?" he asked in surprise.

"You *never* get calls, sounded like a girl... woman. Who is *she*?" his Mother asked, smiling.

"Well… it's complicated," he tried to defer that question.

"Nonsense! It's *never* very complicated, Grant Porter! You cruised through both high school and Purdue as a loner, and now you get a call from someone you gave your phone number to! Who *is* she?" his Mother asked again with a half-smile.

"She's a ceramist, ceramics engineer, Renssalear Poly; she was surprised I could pronounce it properly!" he said, smiling.

"What's her *name*?" his Mother asked, leaning forward in emphasis.

"Amanda McCormick; her mother is a doctor up at Firelands in Sandusky, if you can imagine that! She's a very striking 'Ohia' Girl!" he said with a smile, pronouncing the state as locals did.

His Dad chuckled as his Mother stared at him. "OK, is she worth meeting? I know of a McCormick up there; I think she's a surgeon... she gave a paper in Cleveland a year or so ago. Interesting! Continue, please," she commanded, not willing to wait.

"She's reporting to the NG Materials & Processes... department. I'm starting in a Manufacturing Engineering department. I have no idea how they assign new work or sustaining engineering, if that's the way to say it."

"C'mon, Grant, don't be suddenly *stingy* with words, what's she *like*?" his Mother asked.

"Stand and deliver, Son! She must be something if *you're* at a loss for words!" his Father, Gordon, declared with a big smile, as they both smiled at his sudden shyness.

"Yeah, if she didn't put a spell on you, *describe* her!" Emily, his Mother demanded.

"Well...she's almost my height: tallish, slim...dark brown hair in curls...maybe auburn... fair... big, dark eyes. We had dinner together a couple times. We had langouste at a Chart House on a little marina in Hermosa Beach. They had a good Sangria," he managed to say.

"Oh dear! *She* called you; she must reciprocate some of the interest! You might work together?" his Mother asked.

"Yeah, maybe. I've no idea about how they allocate work," Grant replied.

"So, you said your boss gave you an assignment? What are you starting on?" his Dad asked.

"It's a new die-bonding adhesive, something with diamond dust dispersed within an epoxy film. It's expensive, but it has a *very* high thermal conductivity, so... it allows chips to dissipate higher thermal fluxes. I learned something about hybrid microcircuits at Purdue; I need to get up to speed on thermal conductivity in solids. I think the material of choice *used* to be a copper/tungsten composite."

"It's the usual compromise: copper is a good thermal conductor... about 400 watts per meter-degree Kelvin, but its coefficient of thermal expansion, TCE, is *awful*, almost 20 parts per million per degree C. The semiconductor chips are all down in the 4 to 6 ppm/degree C range; so, if you bond a silicon chip directly onto copper, the copper expands when it heats up and fractures the chip! If you use a copper/tungsten composite 'thermal pad', the TCE is only like about 7 ppm/degree C and it doesn't expand so much. I have no idea what diamond dust's CTE is, but the thermal conductivity is... maybe 1,000 or so. So it's *twice* as thermally conductive as copper/tungsten. That means the chip temperature, and therefore the thermal stress, is maybe half. Diamond-filled epoxy is pretty new," Grant said, thinking about the beauty of using *composite* materials to design in or *engineer* properties.

"I need to get up to speed on conventional silver-filled epoxy films. I don't think they're all that good: maybe... 60-70 watts per meter degree K. The epoxy bonding materials used to be pastes... maybe .002- to .004-inches thick. The film adhesives, usually silver-filled, start around .001-inches thick, so that helps some. I need to get more info! In fact, I might find and reread that chapter in one of my texts upstairs!"

"So she's pretty?" his Mother asked, as if he hadn't changed the subject.

He nodded silently, blank-faced, visualizing Amanda in his mind. "She's stunning; she's... she turns heads in restaurants or Starbucks..." he managed, still nodding.

"This is pretty bad, Gordy! He's *gone*! You need to focus on getting *your* new job started, Grant. Take your time with this Amanda; she may be just an abstraction!" his Mother declared.

"She's a very nice *dis*-traction!" Grant said, smiling. "I'm considering asking her to drive out there with me!" he said, smiling as his Mother rolled her eyes at his Father.

"A couple of dates and you're over the hill! Damn! Adolescence is dangerous!" his Mother said, but she was smiling.

※

Grant sorted some clothes and then went out in the side-yard where his old truck sat in the shade of the garage. He got in and started it up; it started up and idled smoothly. *Maybe* it would survive the drive out to L. A.

He called Amanda about 9:30 from the back porch in the dark; she answered on the second ring. "Hi, Grant! Are you working on your truck or packing?" she asked.

"I'm too tired to do much of either; I'll be up for that tomorrow. The truck might take a while, actually. I got in and started it up; it runs pretty smoothly. That's a good sign. What are you up to?" he asked.

"I did a quick pass through one of my closets; I bagged up some stuff I'll never wear again. Did you tell your parents about us?" she asked. He could almost hear the smile in her voice.

He smiled to himself. "I did; my Mother recognized your mother's name! The... irony of meeting you way out there, is surprising! I enjoyed the lobster and all the rest. We're at the front end of a probably *steep* learning curve. McPage told me I was going to learn to 'drink from a firehose'!" Grant said, smiling.

"My Mother said our *first* job is when we learn *how* to learn. Sometime, you need to explain to me what the *exact* steps in populating a microcircuit are. I've got some of it. Are chips *always* bonded down onto a dielectric substrate?" Amanda asked.

"I'm not sure; sometimes they go down on the package... *floor* for better heat-transfer. This SuperCon film I'm supposed to learn about, helps conduct heat out of the package; I think the package must have to be ... bonded down to some kind of external heatsink. I don't know how they do that in satellites; the structure must be mostly aluminum, but aluminum is just a mediocre thermal conductor: a little less than 200 watts per meter degree K. Copper is twice that at about *400* watts per meter degree K. Copper is too heavy and too soft to work as a structural material. I've got a lot to get up on!" he exclaimed.

"BeO is a little better than Al_2O_3; something like 210 watts per meter degree K. SiC is better than BeO: something like almost 300 watts per meter degree K. I don't know how tough it is to work with; it's really hard and refractory! *Diamond* substrates are being made now; single crystal diamond would be very attractive if you could afford it; its thermal conductivity would be something north of 1,100 watts per meter degree K. I suppose that's why my boss is assigning me this SuperCon stuff; maybe really small bits of pure carbon jack-up the transconductivity. The epoxy bonding films are thin... on the order of a mil....001-inches. I may have to relearn how to do thermal transfer calcs," he finished.

"Hey, do you suppose I need to wear leather shoes out there?" he asked.

She laughed. "I don't know; I saw a lot of running shoes and... sneakers. Do you have any real leather oxfords?" she asked.

"Yeah, I wore them my first year at Purdue; I like the smell of polished leather, but my running shoes are more comfortable and a little lighter," he replied.

"Bring 'em out; you've got plenty of room, don't you?" she asked. "You can wear them if you see other guys wearing them. You don't strike me as a conforming kind of guy!"

"Yeah; I could wear them to formal meetings, I suppose," he replied, smiling at the sound of her voice.

"I don't know about *men*, of course, but a woman can *never* be thin enough, rich enough or have enough clothes!" She laughed. "I like talking to you, or did I say that before?" she asked with a chuckle.

"I do too; I better go in, I'm out on the back porch. I think Mom's listening at the kitchen window; she's trying to gage how... serious this is," he said quietly.

"I've been thinking about commuting with you. That's generous of you; I could try it... put off trying to buy a car for a while. Would that be OK?" she asked.

"Of course; a guy can never save enough money!" he said, making it up!

She chuckled. "OK; get some rest! Tell me how the car *inspection* goes, please!" she said as she rang off.

He woke the next morning in his own bed, disoriented for a moment. After shaving, he went downstairs to find both Mom and Dad waiting breakfast for him. "I called in for a late start; I'm worried about what you're going to find when you look at that old truck closely. Is it *up* to a 2,000-mile trip?" his Mother asked.

"We'll tell you, by-and-by!" his Dad replied, smiling.

They went outside, pulling the truck up in front of the garage door in the shade, opening the hood. His Dad had the truck manual on the workbench; he wrote down the spark-plug number, the air filter number, the oil filter number and examined the tire-description. "I think you might want to run 10W-30 oil out there, Grant: I know it can get up to the high-90's in the summer. Going out through the mid-west it'll be high 80's or low 90's, too. Let's run down to the auto store and mosey around. OK?"

An hour later they were back and pulled the new pickup out of the garage so they could have the older truck on a level slab by the workbench while they worked on the old one. The air filter was easy; the old oil-filter was a bit messy but the new one went on easily. The new oil was easy. The 8 plugs weren't too bad. Grant hadn't changed plugs in a couple years; one of the plugs was badly eroded. "It's a surprise it ran so smoothly with that big a gap!" Gord exclaimed to Grant, holding out the worn plug.

When they had the truck running smoothly again, Gordon drove it down to their local tire store. The tire store mech used a thread-depth gage and just shook his head. "You're not going very far on any of *these* tires; that one's so bald I won't let you drive away on it! Your spare's going on before you leave! I can knock off $75. on a set of 4, but you need 4 *new* tires to drive all the way out to the West Coast in the heat of the damn summer, Grant!"

Grant nodded; the guy was right. It would be risky to try to get *these* tires to L. A.! He'd have to borrow from his Dad, again.

The tire store had 2 of the correct tires but would have another 2 in tomorrow. He'd be ready to head west on Friday morning, if the additional tires came in on time. He had time to do some clothes sorting before supper.

They went home and his Dad went on in to work at noon; his Mom had gone in at 10. He worked through his closet and his dresser sorting clothes, feeling a little relieved at not trying to push nearly-bald tires over the horizon! That wasn't a responsible decision.

When he had two big grocery store paper bags of old shoes and clothes he'd never wear again, he went out on the back porch and thought about calling Amanda.

Would she be home or out shopping for 'enough' clothes?
He guessed that she wasn't as close to broke as he was. He would have to pay this loan back over a couple months; that should be easier to do on his new salary. He took a deep breath and dialed Amanda's phone.
"How bad was it?" she asked, a little breathlessly.
"I'm getting 4 new tires tomorrow, hopefully. The tire shop guy didn't like *any* of my tires! I can pay my Dad back in a month or so from my new salary! The engine's in decent shape after some parts and oil. Hey, this is crazy, but would you like to *drive* out there with me?" he asked before he had time to consider asking it.
There was silence on the line for a long moment. "I suppose we might do that..." she said, slowly. "It would get us out there on Tuesday; 5 days, you said?"
"Yeah, maybe Monday if we didn't run into paving or some other big detour. I can probably do 500 miles per day safely, especially if had someone to talk to, keep me alert! I'm looking at a Rand McNally atlas. I think I'd go west on US 80/90 out to Nebraska, then US 76 to Denver, then US 70 to US 15 coming south out of Salt Lake and down to Las Vegas and down to US 40 at Barstow, California. US 15 runs south all the way to Corona and we're almost to both Orange and Yorba Linda... we might find the

time to look at some apartments before going back to a motel in Redondo or Manhattan Beach. Amy said we could have 2 weeks to lease places." He waited to give her time to think about it.

There was dead silence for a minute. He wondered if he'd offended her. "Grant, that *might* work...can I call you back? How do you get *decent* motel rooms when you plan such a trip?" she asked.

"I don't know... go in and look at the better names when you see them; drive on if you don't like them, I guess. Uh, I think you can actually use your phone or iPad to find them as you go along, Amanda. I know my Mother's done that, going up to Boston, once, to join my Dad for a long weekend. You might be able to make a list of half a dozen top names and just make reservations as we go. We'll get reimbursed for everything, but I don't know how long it'll take Amy to process the expense reports... maybe a couple weeks." Had he been too forward?

The line was silent for a longish minute. "Let me think about it, please. Sounds something like an adventure, actually!" she said.

"It shouldn't be *that* much of an adventure; there must be dozens of good motels out that way! We've done summer vacations when I was younger; there's a lot of scenery going near Denver! Think about it; I'll call you when I've got new tires! OK?"

"OK. Geez, you're *dangerous*, Grant!" she said, but he thought she was interested from the tone of her voice.

He worried about what her parents would say; he worried about what *his* parents would say! They hardly knew each other, but he really liked her company!

He did some vacuuming of the truck cab and hosed-off the rubber and sisal floor mats. The mats dried more quickly than he would have expected in the warm summer air. He got the top of the dash washed and dried; it looked a little better; he sorted through the 'stuff' in the glove box and threw away some of it. He would take the last year's atlas he had perused to construct a route from Ohio to L. A. with him. He hosed-out the truck bed and made sure the tonneau cover worked and locked securely. Once he had tires, he could leave!

Chapter 6   Dinner

Amanda called him at 5:30 PM. "Hi, Grant! I've been thinking about your... suggestion. I suppose we could do it, but my parents may pitch a fit unless we have some kind of meeting so they can get to know you... a little, anyway. Do you suppose we could have *dinner* together, the 6 of us? I don't know how else to do this. If we don't do something like that, it's going to look like we're eloping or something, which we're *not* doing."

She made a kind of sense with that. "It would maybe get my Mom calmed down. She... *grilled* me about you this morning at breakfast! Dad was less intense. Yeah, maybe dinner somewhere... neutral territory might work. Where should we go? We don't eat out much, Mom usually cooks something when I'm home."

"I don't know. How 'bout Erie Lodge out on Catawba Island? It's got a nice menu on-line. We ate there a couple years back at a birthday; it's got a big porch overlooking the lake and the ferry terminal."

"That sounds OK; we're something like 45 minutes away. That's not too bad," he decided. He could do dinner with her parents, he hoped.

"You'd like to head west on Friday if you get tires on Thursday?" she asked.

"Yes. I think we should report in and get started working; we start accruing vacation days after 6 weeks, if I read the legalese on the offer we signed," he said.

"I read that, too. We'll get a long weekend on the 4th and again at Labor Day. We could do something out there, I suppose," she suggested.

"There's going to be a *lot* to do out there, Amanda: the Golden State has lots of day parks and places to hike. My Dad had a book, out of date but interesting, about driving north on the old *Camino Real*, the 'royal road' from San Diego up to San Francisco. It follows California Highway 1, Pacific Coast Highway or PCH; it has commemorative signs in lots of places. We might find a newer version," he suggested, smiling in anticipation.

"We're going to have to learn some *Spanish*, Grant. I had 2 years of Spanish and French in high school; I can cope!" she said with a chuckle.

"We might need to learn some Chinese or Tagalog or Vietnamese! There are a *lot* of Asians in the Hybrid Assembly area; there's one family of 7 sisters that Amy mentioned in passing. They all work in the one big assembly room!" he replied.

"I met a couple of Taiwanese and a Japanese guy up in R6. There are a lot of Asians at NG; they are probably very technical," she said.

They agreed to push for dinner tonight, so he left a voice mail for his Mother with the request. He managed to reach his Dad. "Dinner, huh; Emily might like that a lot! I"m OK, tell Em I agreed!" Gordon said.

Grant went outside and sat in the truck, thinking about the old truck, Amanda and his new job in a succession of thoughts, none of which fit into any kind of logical arrangement. Driving out to a new job with Amanda would certainly give them time to know each other a little bit. He wondered about her hobbies and interests.

His phone rang: "how'd your parents respond?" Amanda asked.

"Well, Dad's in; I left a voice message for Mom. I may not hear from her for a while. Is the idea of the two of us driving out there together completely crazy? I'm looking forward to it, but it might be a stretch." he invited.

He could hear her breathing as she considered. "It's a stretch, as you say; it'll give us time to get to know each other. I can't tell if Mom's offended or interested. My Dad laughed when I managed to tell them after dinner. We're... of age and all. I think it's an interesting... stretch. We're going to need to make an agreement about a couple of things, I think," she said, with a sigh.

"Separate rooms, for starters," he replied. "We can do that; I... uh...I'm having trouble talking about this, even to myself. I know engineers often travel in small teams on technical trips; if they can do it, we can do it, can't we?" he asked rhetorically.

"I think we can. I'm attracted to you because we can talk so easily... I never had anyone my age to talk to! I like the idea of sharing ideas with you... a lot! You see things I never thought of: lobsters and canoes and... airplanes, I suppose. It would be *nice* to have my... *our* parents accept that we're going to do our own thing at some point. My Dad seems to be more OK with it than my Mom; I didn't expect that," she said slowly.

"Is that a male thing: more easy going?" he teased.

"Yeah, us women are higher strung and all," she said sarcastically. They both laughed.

"So, we'll meet up at Catawba at the Erie Lodge at like 6:30 or 7:00 depending on when we can get all the troops into a car. OK? I'll call you once I talk to my Mother. She's usually home by 6; I'll call if there's a problem!"

"All right; hey, you didn't talk about it, but I was thinking that *building* an airplane, home-building an airplane, would be a really good way for an engineer to learn a bunch of things. Did *you* ever want to do that?" she asked.

"Yes, I *did*! We just never had the money! Dad's friend, Ernie Newton, the jeweler I worked for, has a Cessna 172. That's a 'general aviation' airplane, factory built and certified by the Feds to meet certain standards. I wanted to build an airplane a lot smaller and more agile than that *truck* of a plane!

Experimental homebuilt aircraft are often kit-built by individuals; that's a Federal category of aircraft. The builder becomes the mechanic for that aircraft. I've looked at several of them; I flew in one once for about half an hour; that was eye-opening! They are a *lot* more fun to fly!"

"A C-172 weighs something like 1,700 pounds empty; max weight at takeoff is something like 2,300 pounds. *Experimental* aircraft, not certified, can be single-seaters with empty weights of as little as 400 pounds up to 4-seaters with empty weights of 1,800 pounds! There are dozens, maybe a couple hundred, experimental designs that actually get built and fly. We can talk about them on the way. Yes, there's a lot of technical details in designing and building a small, light airplane. A couple of the newest designs are powered by a single electric motor driven by lithium batteries. By the time I have the money to build something, I might just go for an electric propulsion system!" he said with a chuckle.

"*That's* what I meant: you can answer questions so easily! Changing subjects: do you have any notions of why M & P *isn't* doing the introduction of diamond-filled adhesive films? Why is Manufacturing Engineering doing that?" she asked.

"I've no clue, Amanda; it may have to do with money or 'intramural' politics... I don't have any insight into how those different tasks get assigned; we both have so much to learn about the culture at NG! We don't even know *who* to ask that question of! Ask your boss as soon as we get started! He had better know the answer; I'll ask mine. She had better know the answer. I was struck with how knowledgable McPage was. He *has* to have some kind of tie-in to the technical stuff or he wouldn't know anything about who to assign us to!"

"Yeah, I noted that. HR must be tied into 'engineering' fairly tightly to be able to predict what we'd be working on. It'll be interesting to see what My-An, Jeff and Will get assigned to. I suppose we'll get that info when we go to our first section meeting. I'm going to want to learn what you're learning... is that OK?" she asked.

"Of course. That's the whole point of having a technical staff; they have to *share* some mutual knowledge of how things work! In a sense, they need to 'train' and support each other, don't they?"

He chuckled. "We going to do that? Train each other?"

She laughed: "you need some *training*?"

"*Maybe*; we might want to consider that, I suppose!" he suggested with some implied innuendo!

"One supposes we'll 'get around to' that, Bub!" she said. He had a round 'to it' somewhere upstairs! A yellow, coin-shaped plastic token with 'to it' embossed on one face. She might get a kick out of seeing that!

The Porters arrived at the restaurant at 6:45 PM. His Mother was very interested to meet this Amanda person his son had obviously fallen for!

Amanda introduced her parents, Gale and Marilou, and then Grant introduced his parents standing on the broad porch of the old lake-shore resort. They went inside and were escorted to a table by the long windows that faced north, looking toward South Bass Island in Lake Erie or 'Put-In-Bay', where Perry's 1812 Victory Monument stood.

"I thought it was a good idea for us to meet; we have some things in common and a couple things we might discuss. Grant and I are professionally involved now we're both going to work for Northrop Grumman in Redondo Beach. The campus was originally TRW's Space Park; we're probably going to be involved in helping build satellite comm systems," Amanda explained.

"We're getting to know each other; we have some common tasks to get completed. We have to get our 'stuff' out there, rent apartments, transfer banks and get registered to vote. I have to get California plates and insurance for the pickup. We have to get 'registered' or something for medical and dental services, and undergo some kind of security investigation. Our contacts in HR help us do all that."

"I asked Amanda to ride out with me because I think she has more stuff than I do and I think I have plenty of room so she won't have to ship it out there. Dad and I think the pickup will be ready for the drive after we get new tires on it. I got the air conditioning system recharged today," Grant said as he watched Amanda blush just a little.

Marilou, *Dr.* McCormick, Grant reminded himself, shook her head. "This is worse than I thought, *she* starts and *he* finishes!"

"You seem very comfortable speaking about the two of you. How'd you manage that? You've only known each other for a week!" Emily asked, looking at Amanda.

"Well, we're the same age, about; we have some interests in common; we're liable to work together sometime. He's a materials engineer and I'm a ceramics engineer: our specialties, if we have any yet, are going to overlap at least a little," Amanda replied and swallowed.

"I never had anyone my age to talk to except for Kitty, whose mental age is 3 years behind her social age, and knows very little about anything except clothes and cheap jewelry. Grant knows a lot more about real jewelry; I'm already learning some new words. We talked about flying; I don't know anyone that's a pilot. I might enjoy learning how to do that, sometime!" Amanda finished.

"You expect to spend 4 or 5 days going out," Gale asked.

"Yeah, if we spell each other, we'll stay more alert and be a bit safer. I've got a route selected that we'll modify if we hit bad weather or roadwork. On flat ground, like Ohio, I get about 19 miles per gallon; I don't know what to expect when we go through Denver, but we should get 500-plus miles per day, or about 9-something hours driving in daylight, each day. I don't plan to drive at night or go more than about 60 miles per hour," Grant explained.

There was a short period where they ordered drinks and salads. Grant ordered diet cola; Amanda ordered iced-tea, along with her Mother and Emily.

Grant decided on a small filet mignon with a side of succotash and a baked potato. They got three small cutting boards with small, fresh-baked loaves of bread.

Dr. Marilou and Emily knew other people who worked in their 2-county area and found it easy to talk 'shop'. Gale McCormick had worked his way up in the Sandusky police department before running for councilman. Gordon told stories about selling 'heating equipment' all across the Mid-West and up into New England as one of a senior salesman for an industrial furnace supplier.

"So what are you going to be doing once you start," Gale asked Grant.

"I met my new boss, Paula Hopkins, for at least 3-minutes, right after we signed our employment agreements. She wants me to see about introducing a new, high conductivity epoxy bonding compound; it's something called *SuperCon.* It uses diamond dust to increase transconductivity in hybrid microcircuits. There's some reason to expect that Amanda may become involved with it, too," Grant explained.

Amanda said: "I met my new boss briefly, too; he had his secretary run me all over the building meeting some of, hopefully, most of, my engineering department people. I don't know what he's got slotted for me, yet."

Amanda made eye-contact with Grant often, usually smiling as they ate and talked. The steak was pretty good; thick enough to be rare in the center. The restaurant used butter and something with tarragon in the sauce.

Amanda had ordered a pork dish that she seemed to like; Grant could tell she was a bit nervous trying to answer questions as they went. She lifted her chin once to him. He finally figured she wanted to meet him back by the restrooms.

He used his napkin and excused himself, trying not to smile visibly. Families were suddenly complicated!

She met him in the corridor leading to the restrooms and pointed out a door to the porch. "Damn, *this* is stressful!" she said with a smile. On an impulse, he gave her a quick kiss on the lips! She grabbed his hands. "I *needed* that! Your sense of timing is very good! Save some of that, please! We're not going to get any time alone until we get in the truck and go!" she said in a soft voice.

"Hey, it's going OK; they're ignoring the question of sex, which we don't want to talk about!"

She nodded, looking into his eyes. "Back into the fray!" and they went back to the table. Grant passed on the dessert and had a mediocre espresso instead.

"Not good?" his Father asked, seeing the look on his face.

"So-so; not hot or rich enough, my mokapot makes better stuff!" Grant replied.

"What's a mokapot?" Amanda asked.

"It's one of those aluminum castings that boils water up through ground espresso beans. They're used all over Europe. All I need is a stove top, water and my ground beans. You'll get to taste it soon; its as good or better than most espresso shops make!"

"*That's* a challenge!" Dr. Marilou declared. Grant nodded to that.

Chapter 7 Heading West

Grant pulled up at Amanda's white clapboard, two-story house; it was bigger than he expected with something like 3 or 4 gabled bedrooms upstairs, he guessed. He opened the tonneau cover and lowered the tailgate.

Amanda and her Father carried out 3 suitcases, a wardrobe bag and then went back to get a series of packing boxes, some of which had labels from RPI on them. Grant pointed to the cooler which had to ride in the bed against the tailgate so they could get at it easily. "We need to anticipate how fast we're going to drink this stuff: water, apple juice and diet cola. You need to tell me what you want, please, Amanda!"

"I'll drink water to start with; the apple juice will work, too!" Amanda responded, smiling at him nicely.

After saying goodbye to their cat and her Father; Grant shook hands with Gale. "Be careful out there, Grant! It's not the country it used to be!" Gale said.

"Yes; it's all at cross currents with itself, I know!" Grant said. We'll tell you where we are as we go!"

Amanda kissed her Father on the cheek and got in the driver's side and they were off. "Oh my God, that was *hard*!" Amanda declared, leaning over and kissed Grant on the cheek as they went around the first corner. "I owed you that for surprising me at dinner!" she said with a big smile.

"Are you OK with us, for the moment at least?" he asked.

"Yeah, I am; I can relax now. It's very hard to *not* talk about sex; Mom's assumed we're having sex! We had a discussion about contraceptives! I'm not offering, but it might be easier! I'm going to have to *not* talk about it until sometime... later. *Damn!*" she was a little angry, he decided. "You get any of that?" she asked.

"No; it was there in the room like a 600-pound gorilla, but it didn't get to speak!" he chuckled. "The scary thing is, we *know* they went through this! They still can't talk about it!" Grant said.

"I need some science and less emotion: she's got to let me grow up! What's our first way-point?" she asked.

"We're out of Sandusky for Bloomingville and the Ohia Turnpike; Perrysburg is about 60 miles. You want to switch there?" he asked, enjoying looking at her in quick glances, sitting beside him in a tailored shirt of some kind and slacks.

"Yeah, that'll work. What's after Perrysburg?" she asked.

"The Indiana Toll Road is about 4 hours or so; I highlighted the route in the various pages of that atlas. We can just follow along!" he suggested, pointing to where it lay open on the seat between them.

"Stay this side of South Bend, maybe? Mom showed me an app for finding motels. I hope it works!" she said, thumb-typing on her iPhone. A moment later she sighed. "This is going to be good, Grant! The dinner worked pretty well, don't you think?"

"I do and it did. It didn't leave either of them much to say, since we can't talk about sex comfortably! I... that bothers me, but I'm surely not going to try to talk to *my* parents about it!" he sighed with a chuckle. She laughed, nodding.

※

They stopped to get lunch just outside of Bristol, Indiana at a McD's; it wasn't wonderful, but the chicken nuggets were at least freshly deep-fried. With the different sauces, they were pretty good.

"You keep your truck pretty clean! I like that. You're pretty neat?" she asked, studying his profile in glances while she drove. He handed her pieces of chicken with sauce already on them so she could concentrate on driving.

"Yeah, I like to keep things neat and clean; I suppose I'm a...'*neatnik*'," he replied.

"I can do that! It's easier to find things!" she replied.

They switched drivers again and she experimented with her iPhone and the new app. "There's a Hilton near where Indiana 31 crosses the Toll Road north of South Bend; it has good reviews. Want to try that? We can do an early stop the first night," she asked.

"Sure! Gotta start someplace. We might go get rooms, clean up and go find dinner? Should be less stressful than the Lake House!" he said, teasing.

"Roger that! I think *us* calling the shots on a first dinner took a lot of emotional baggage off the table. Mom was pretty mellow afterwards; she asked me what some of your other interests were. I didn't press her for any kind of assessment; I told her I'd discuss *us* when we were out west. Dad smiled while I was trying to tell her we'd be 'mature', or something; he's been better about letting me do my thing than she has. Did your Mother say anything?"

Grant cleared his throat and rolled his eyes. "She said you: 'were very impressive... very composed and very pretty'! And she gave me some ... *stuff* about being a loner for so long she thought I was *never* going to pick someone to be with! I think she was relieved that we'd run into each other. You OK with that?" he asked.

Amanda nodded, a small smile on her lips. "Yeah, she's pretty *composed*; that's a good word. I suppose women doctors work at being composed, working with people who are in pain or in fear of being in pain. I don't think I could *ever* do what my Mom does."

"I'll take plain old *things* every time! She had a *cow* when I wanted a quad-rotor drone back a couple years ago. She thought I was weird and wondered if the 'other girls' were into drones. I told her the 'other *boys'* were! I took a bunch of neat aerial pic's around our neighborhood... terrorized the cat. Chased her away from a sick Redstart, once! I don't know if the bird survived, but at least *she* didn't kill it! You ever get into drones?" she asked.

"I didn't; I was busy at school and then at the jewelry shop. I... like the idea of *flying* a real airplane, but we couldn't afford it," he replied.

"I got into side-by-side shotguns when I was a senior at W. W. Ross. Dad goes hunting with his buddies every Fall; one of them had this beautiful Merkel hammer gun, a side-by-side 12 gage! *God*, it's a beautiful piece of art! It's... engraved, the receiver, the hammers and the lever. Most of his friends shot Winchester Model 12's and Remington 1100's: really plain 'meat guns' with *no* character except they work," he tried to explain.

"We found a 20-gage AYA side-by-side from Spain in a small gun shop down in Tiffin; the guy who owned it got it from some relative; he wanted something that shot three rounds fast, so he wanted just enough money to buy a Remington 1100 in 12 gage. He had a piece that was worth... maybe $2,400 and he *wanted* something that cost maybe $450! I borrowed it from Dad and paid it back in a couple months! That gun is in a gun-sock in my closet, waiting for later. I may bring it out to California; they hunt a lot of quail out there, I guess. I might want to learn how to do that sometime. You ever hunt?" he asked.

"Shotguns, huh? I never paid much attention to them. Dad doesn't hunt anymore; he has weak ankles. He taught me to shoot .22's when I was little, maybe 9 or 10. I can do pretty well off-hand up to about 40-50 feet. I haven't shot anything in a year or so. Shooting is such a *polarizing* thing; some people wax eloquent about dropping a duck out of the sky on a cold and snowy pond; others, mostly girls and women, just don't see anything *interesting* about punching holes in bottles or cans or paper. I gave up talking about it because some of my friends thought I was some kind of NRA idiot!" she said, shaking her head. "You hunt pheasant and ducks, right? Most Ohia guys do," she asked, smiling at him.

"Yeah, when I'm home and can hunt with Dad; he can get us into several places, farms, where they have enough pheasants around to let their friends shoot some. A year ago we shot quail on one farm; that was different! A quail is a lot smaller than a cock-bird! That little AYA worked very well for me: I even managed to get a double! Surprised Dad when the little covey broke and I got two of them: bam! bam!: just like I'd done it before! We had quail and pheasant that night; I'd never had quail before. Read about it, heard about it, but never had 'Gentleman Bob' rise up in front of me before!" Grant related.

"Gentleman Bob?" Amanda asked.

"Oh, Bobwhite Quail are smallish, not much bigger than a robin. They tend to hold under certain conditions unless they're hunted a lot. That means that even without dogs they'll sometimes *hold* until you're only 20-30 feet from them. Out in the West, the quail are a little bigger and in bigger flocks and they run like crazy! So Western quail are *not* gentlemanly and sometimes you have to hunt them on the run to get near enough to shoot. They almost *never* hold unless you have dogs! There's also a big quail out there, the Mountain Quail, and he's up in the steepest terrain he can find because he's almost never hunted there!" Grant explained enthusiastically.

"You're a kind of 'wikipedia', Grant! I don't think my Dad hunted quail in Ohio; he talked about a trip he made as a young man to a plantation in South Carolina. He raved about shooting doubles in some kind of grass... marsh grass. Does that sound right?" she asked.

"Yeah, there are lots of private hunting reserves in the Southern pinelands; those are all Bobwhite Quail. Someday, if I'm ever rich enough, I might just *pay* to go hunt lazy Bobwhites!"

The conversation flowed as he drove and she guided him using her new app to get to the Hilton where she'd reserved two rooms for the night.

They went into the lobby where she asked to see her room, and after a quick inspection, they registered separately, getting looks from the desk clerks as they wheeled roller bags to two rooms.

Grant took a quick shower and changed into clean clothes; they asked at the desk for a good local restaurant. The clerks gave them a choice of 2 places, both just beyond walking distance. They settled on The Old Plantation, an up-graded version of a Southern, white-pillared plantation estate.

The service was a little slow but the prime rib was pretty good! Grant was pleased Amanda would eat pretty much everything he would; that was 'comfortable', looking forward. He didn't know exactly what to expect from her; she was very different from the few girls he'd known in high school.

They went to her room afterwards and scanned through the road atlas to confirm tomorrow's route. Amanda called her Mom to 'check-in' and keep her apprised of where they were. "The truck just hums and the A/C works fine; we're very comfortable; there's a lot more foot-room than on a jet!" she told her Mother.

It was a little awkward when Grant decided he needed to get some sleep. They embraced, a little awkwardly and he went down the hall to his room. He thought about the embrace for a while before falling asleep. She had had no aversion to a very nice kiss!

In the morning, she called him while he was shaving. They went into the motel's breakfast room and had orange juice, scrambled eggs, buttered toast and he tried a weak 'urn-coffee' while she had tea. A few minutes later they were checked out and in the truck they headed West on the Indiana Toll Road.

He opted to drive as they entered the 'belly of the beast' in South Chicago: Gary to Joliet was crowded and busy! Grant couldn't tell if today was normal or heavy because he'd never driven that stretch of highway before. The traffic was sometimes 5-lanes each way, almost bumper to bumper at 45 miles an hour! Leaving Minooka, Illinois, was a relief as the traffic thinned out and he could stay 100 feet behind the car ahead. They stopped and switched drivers; Amanda drove parallel to the Illinois River in the stretch before reaching LaSalle.

She pulled off the interstate to go to the bathroom at a gas station. He pulled the truck on up to a gas pump and topped off the tank, a bit nervous as 3 noisy bikers pulled in under the roof to gas their bikes. They wore common black leather vests over no shirts, exposing dirty, tattooed flesh: looking like Harley outlaws in German army-like helmets. Grant was immediately pissed but worried; there were 3 of them; he was outnumbered in any confrontation!

One of the bikers had an excess of bad attitude and smiled when he saw Amanda coming back to the truck. "Hey, doll; I got something you need to try!" he called loudly as his friends smiled, watching this 'sport'.

Grant had the tonneau cover up to get drinks. He got his hand around a 36-inch length of half-inch diameter pipe, the only weapon he had. As the biker walked toward Amanda, waving a big hunting knife out to the side, Grant stepped around to the guy's right side, the pipe held low, down beside his leg.

When the biker turned to confront him with the knife, Grant rushed him, jabbing him in the throat with the end of the pipe, a move that hid the nature of the threat! The fencing stop-thrust caught the fat biker in the upper chest or throat and stopped him in his tracks, knocking the steel helmet off his head, partially gagging him; when he tried to raise the knife with his right hand, Grant put all his strength into an overhead coupe, crushing the man's right elbow with the sound of meat being crushed and bones breaking! The man was driven down to his knees and collapsed on his side as Grant stepped forward to follow up.

"OK, asshole, go for the knife so I can take your filthy head off!" His voice was hoarse, very angry, the pipe poised in the air for a killing blow to his head.

"Grant, stay out of my sightline!" Amanda ordered, legs spread, both hands in a Weaver stance, black pistol pointed at the taller, larger biker whose hand hovered over a pistol in his belt, under his vest.

Grant goggled and then smiled; she had him dead to rights where he stood!

The second biker, tall, bearded and heavyset, hesitated and then started to draw his Colt .45 out of his belt. "You're not going to shoot me, Honey Doll..." and she *did*! Three aimed shots rang out, echoing off the metal roof: the first hit his right hand, knocking the bigger pistol out of it; the second hit his right elbow, turning him partially away from her; the third drove into his left shoulder, knocking him down to his knees. "Next one's a beauty-mark, Cockroach!" Amana barked.

The big biker went into shock about that time and sagged down, trying to hold both wounded arms in his good hand as he lost consciousness and fell over on his side.

Moments later a police siren wailed coming off the Interstate; the station operator had called the police! A few moments later an Illinois State Trooper got out of his patrol cruiser and walked forward, a billy club in his right hand. He used the billy club to touch his hat to Amanda who held her stance, gun trained on the third biker who was down on his knees, hands clasped on his head, crying in fear that he'd be next. The Trooper sighed as he studied the scene in silence.

A swarthy proprietor, maybe Asian Indian, came running out of the store and told the Trooper in very rapid words that the 3 bikers had attacked the young couple. Grant couldn't be sure, but maybe the Indian had been troubled by this trio before, or knew of them. He started to relax as the Trooper nodded, studying the scene.

Grant walked slowly over to Amanda; she'd left her sunglasses in the truck; her face was pale and angry, as she stood there, trying to recover her composure. She was angry, too!

The Trooper chuckled and shook his head; "young lady, either you're *very* good or very lucky. That was damn good shooting! I *do* hope you've got a permit for that Glock!"

Amanda nodded and relaxed her stance, still holding the pistol in both hands, but it now pointed up at the roof above them. The Trooper used the billy club to move the .45 away from the second biker who lay on his side, his injured right arm bleeding on the blacktop, probably going into shock from 3 bullet wounds.

Amanda switched magazines and the pistol vanished into her purse, which Grant now recognized as one with a holster sewn into the back side. He had no idea she was armed and carrying! She held out a card to the Trooper as he came up, smiling, his flat-hat pushed up so he could see in the shade of the gas station roof.

He scanned the card and her face and turned to Grant. "You know what they say about bringing a knife to a gun fight, don't you, Son?"

"Yeah, but *we* didn't pick this fight and I had no idea she was carrying! I was responding to this asshole's big, shiny knife when he threatened her! He was the initial threat... I had to stop *him* first," Grant replied, pointing at the knife on the pavement, his voice a little hoarse with rage, the pipe down at his side again.

Amanda handed the Trooper a second card, a Sandusky Ohio 'courtesy card'. "My Father's a city councilor, a former police detective. I don't know if this is of any value here, but he told me to always show it to a law officer."

The Trooper spoke into his shoulder mike for a minute, mostly numbers and then asked if he could examine her weapon. She pulled the Glock 17 out, ejected the clip, catching it, shucked the action, catching the cartridge from the chamber and handed it grip foremost to the Trooper, who was still smiling.

"Impressive, Ma'am; someone trained you very well! 'Amanda McCormick' of Sandusky Ohio up on Lake Erie by Cedar Point!"

"My Dad taught me to shoot years ago!" she said, still angry and maybe a little bit scared, now.

"You tell him I said he trained you well! Justified shooting! OK! I've got to do a report and you both have to sign it, but I don't think these guys are going to object to a trip to a hospital and then to jail for assault with deadly weapons with intent to injure or rob."

"You did *that* with a pipe?" the Trooper asked as they watched the fat, first biker vomit and try to sit up to hold his broken arm, now bent oddly out of place, before he fell back on side.

"I... don't carry... this was the only thing at hand. I took a knife out of a guy's hand once before when I was still in high school. It works pretty well if you're fast enough!" Grant explained, trying to get his rage and his pulse under control.

Amanda was coming out of her focus as the Trooper checked the serial number on the carry permit and handed the weapon back to her. "You know to clean that as soon as you can, right?" he asked.

"My Dad taught me to do that years ago, Officer!" she replied in a very soft voice, not quite shaking.

An ambulance arrived with another State Trooper escort. Two EMS techs in white, came over to see the injured bikers. "Do we *have* to try to save these arms?" the one EMS guy asked. "Man! They look like real upstanding citizens! Looks like they picked the wrong people to mess with! Geez, that arm is *really* wracked up; with a little bit of luck, he'll lose it and be unable to ride! Good work!"

When the Trooper asked where the two of them were going, Grant explained they were driving out to Redondo Beach to start work at NG as engineers.

"New jobs?" Trooper Erickson asked.

"First jobs on graduation; she's from Renssalear Poly and I'm from Purdue," Grant explained while Amanda stood in silence, recovering.

"Good luck; keep your eyes open!" and he waved them away as the EMTs loaded the two injured bikers into their van, and the third biker went into the patrol cruiser.

Chapter 8   Reset

It was another hour before Grant pulled out of the gas station after paying for his gas. Amanda had a bottle of orange soda and a hooded look as she drank it in silence, turned towards him in the passenger seat-belt.

"Are you OK, Amanda? Please talk to me!" Grant asked as he idled along at only 50 miles per hour, head turned to study her. She was in a kind of emotional shock at the attack and having to shoot someone.

"I'm OK... that was scary... never shot anyone...Gotta call Dad, I suppose. Hey, you were amazing, that fat jerk never knew what hit him, did he?"

"I fenced some in high school and freshman year at Purdue. Fencing moves are designed to be unstoppable, well, saber anyway. The stop-thrust to his neck is almost invisible because the pipe was aimed at him: he couldn't see it. When he raised the knife, I chopped down... broke his elbow pretty good, I think. I don't think he'll be riding anytime soon! If he loses the arm, he can tell others how stupid he was! Damn!" Grant exclaimed, still angry but starting to feel better that it was over.

"That was too close, *Grant*! It could have gone very badly... you were an easy target if I hadn't been armed and ready!" Amanda said, frowning.

"Yeah, but you *were*; you were *beautiful*! That was *very* nice shooting, Amanda! I had to stop the first one; I assumed the tall one was going to draw a pistol; I thought I'd have to throw the pipe like a boomerang and hoped I could hit him hard enough to slow him down, when you told me to stay back so you could shoot! Wonderful words... I didn't know you had a permit to carry! When were you going to tell me? 'Cop's daughter shoots scumbag!' I can see the headlines! Damn! Damn! *Damn!*" Grant was up on adrenalin, but he could see Amanda replaying the attack in her mind, worried about might have happened.

Something in Amanda eased and she smiled, some color coming back to her cheeks. She teared up for a moment. "Shouldn't have to *do* that!" she said, putting her arm around Grant's shoulders and hugging him. "Those ugly animals shouldn't be out on public roads! I considered shooting him in the forehead!" she murmured. "I don't know how close I came to that... then I'd be a *murderer*! I don't know if I could live with that!" she sighed.

"Those two animals got exactly what they deserved! Maybe I should try to see if *I* can get a permit to carry! What a blow to my fragile male ego to see you do that so easily, like you did it every day!" He smiled, they had been skillful and lucky. She was too threatened at the moment for humor.

"Hey, you're going to need to stash that Glock in the truck out in Redondo Beach. One of the clauses talks about no weapons in any of the buildings!" Grant remembered.

"Yeah, I thought of that. I carried at RPI after one of the rapes... never saw anything I needed it for up there. *Damn*! That was too close! I need time to think about things. I better see if I can call Dad!" she said, almost to herself.

Amanda called her Father while Grant drove. He stopped so they could get more water from the cooler, while she explained to her Father what had happened.

Grant drove on west on US 80 towards Moline and the Iowa State Line at the Mississippi; he was thirsty and drank another bottle of water. The second night they chose a Ramada Inn just east of Omaha, Nebraska, not far from the Missouri River, having lost a bit more than 2 hours in the attack at the gas station.

Amanda was still subdued but not completely quiet after she talked to her Father. Her Mother called a little later as they pulled into the motel. She talked quietly in a corner of the lobby as Grant signed them into separate rooms again. Amanda wanted to clean up before looking for dinner; she took a longer time to get ready for dinner this evening, very distracted.

They walked to a nearby steakhouse and went inside; the very young-looking waiter led them to a table by a wood fireplace not needed in the early summer's heat, here on the shortgrass prairie.

Grant ordered a small ribeye; Amanda ordered a salad and iced tea. The restaurant didn't have espresso, so Grant ordered diet cola. He studied her as she sat still, reliving the attack. He sent his Dad an email telling him to call Gale, Amanda's Dad, for the story of the attack, emphasizing they were both unharmed and OK. He said he'd call home tomorrow.

After dinner they walked back to the motel as dusk fell; Amanda was still subdued and maybe melancholy about the attack. He walked her to her room and said goodnight, before walking to his room, worried at how upset she was. He decided she needed time to sort things out and went into his room and lay down, the drapes open so he could see the amber sunset to the West, reflected off a cloud bank in the sky. He wasn't in a daze, but he was subdued too, brooding about the attack; they shouldn't have had to go through *that*!

He was jarred out of his reverie when his cell phone rang. "Grant, can I ask you to come over here? I don't want to be alone right now" Amanda asked.

He guessed she'd never faced violence before. He left one light on and walked over to her room. She let him in, a little downcast. "I need a hug right now." They embraced, standing awkwardly by the TV across from the bed. After a minute or so, she asked: "would you stay here with me, please? I don't think I can... be alone tonight."

He didn't see any options; he would try to be her refuge for the night. He decided not to go back to his room to change into his shorty pajamas. He would sleep in a tee shirt and his underpants. She went into the bathroom, brushed her teeth and emerged in flannel pajamas. She left a light on in the bathroom to provide some light in the small room.

"I know this is... awkward..." she started to say.

"It's OK, Amanda; we managed to protect ourselves today. I think we need some time to unwind a bit. We can take a time-out for one night," he said

She got into bed under a sheet; he awkwardly lay on top of the sheet beside her, not sure what he should or shouldn't do. He got part of a light coverlet up over his legs and tried to relax lying beside her, very aware of her presence. He could smell her fragrance; this was *awkward*! He lay there, unable to relax in her proximity. She turned towards him and put an arm around his shoulders, in silence, a bit awkwardly, catching one of his hands in hers: big eyes wide, staring, traumatized.

He lay there, eyes closed, sure that he wasn't going to get much sleep while she was reliving the attack!

He fell asleep despite his misgivings. He woke later, sometime after midnight, rolled on his left side, away from her. He fell asleep again, emotionally drained from fear and anger. He woke a little later when she had gotten her right arm over his shoulders, still asleep, hearing her very soft breathing as she slept.

He woke when the rising sun sent a shaft of light into the crack between the curtains in her room. He lay there, surprised he'd made it through the night. He had to go to the bathroom! He wanted her to sleep and tried to slide out off the bed slowly; it woke her up abruptly.

"I... thank you for staying with me; I'm OK, I think. I know I acted like a dork, but I might have killed that jerk!" she said, yawning and rubbing one eye.

"That jerk might have shot *me*!" he replied. "I chose the knife threat that was closer to you; I'd have never made it close enough to the second guy in time to keep from getting shot! Thank you for shooting him! I think the Trooper was most impressed with your skill! I'll talk to your Father sometime about yesterday. We're they OK when you talked to them?" Grant asked.

"Dad was fine; Mom was worried. I told her we were fine, unharmed. I think Dad's going to go through police channels to get more information from the Trooper who booked them!" she said.

"Amanda, I gotta go to my room and shave. OK? I'll be up in 20 minutes. You OK to go get something to eat? We need to get back on the road!" he asked, hoping she'd had time enough to recover.

Half an hour later they pulled into a Cracker Barrel Inn for breakfast after they crossed the 'wide Missouri' River. The service was surprisingly good; she ordered an omelette with toast and tea; he ordered a scrambled breakfast in a cast-iron skillet. The coffee was hot, if not very rich; he needed a Starbucks!

They got back in the truck and another 20 minutes later, she guided him to a Starbucks in an Omaha suburb using her phone.

He drove on West towards Lincoln with a quad espresso machiatto, much happier. Amanda drank a double espresso and seemed less melancholy in the early morning. Interstate 80 was less crowded as they went on west towards Grand Island, the sun climbing up the sky behind them. She asked about the incident he'd referred to back in high school where he'd used a pipe to disarm someone.

"That would be Donny Black!" Grant said with a small smile. "We were seniors. He was impressed with himself as a running back on the football team. He wasn't very big or very fast, but he thought he was; he was... a bully. He harassed me one morning going into gym... guy crap, full of testosterone and no brains. He grabbed a piece of railing... wood... maybe 6 feet long and said he was going to stick it up my lilly-white ass! Pissed me off for some reason! I backed away until I found a piece of steam-pipe that a janitor had left leaning up against one of the hall lockers. I parried his first stroke pretty well, knocking the wooden pole almost out of his hand. I charged once I was inside the pole and slid the pipe down the pole until it hit his fingers, he yelped and tried to back up but I was inside his weapon and I decided to hurt him! I don't know why. I poked him in the chest with the end of the pipe hard enough he needed stitches!"

"He was surprised again when I knocked him down with my knee in his crotch; I put the pipe about a foot above his eyes and said I was going to drive it through his head into the concrete floor! The janitors pulled me off him about that time! He was booted off the team and never did get to play. He avoided me every time we passed in the hall! It taught me that... *sometimes*, an attack is safer than a retreat!"

Amanda smiled and shook her head. "Boys, men, are *so* physical! Girls almost *never* get physical, however upset they get. Dad swore *women* were better police officers because they *didn't* want to smash somebody in the face! You were angry yesterday, weren't you?"

"I was! I was outnumbered and I was pretty sure one of those bikers would have a gun! I could take the first guy but I had no idea how I was going to take the second guy! I was afraid he'd shoot me! It sure worked out! *Damn*, was I very glad to hear you tell me to stay back: I was dumbfounded for a second there!" Grant said, grabbing her hand with his.

"We worked very well together, actually! You surprised them by attacking so quickly; the tall jerk didn't believe I'd shoot him... so I shot him 3 times to disable him. I couldn't *see* where the bullets went! I expected to be able to *see* them hit him! It puzzled me, and then I saw the blood. Shouldn't that many shots have knocked him off his feet? I expected him to fall over! I was afraid that he wouldn't go down!"

Grant smiled. "What did your Dad carry? I'll bet it wasn't a 9 mm!"

"He carried a customized .45 Colt; he said it was too big for my hands... said I couldn't control the recoil. He suggested either a .357 revolver or a 9 mm semi-auto. I think the 9 mm is faster," she replied. He nodded; it was.

"If *I* were to carry, it would be a .45; it's a big, slow bullet, but it strikes with a *lot* more momentum than any 9 mm! Dad carried one in Europe; he knocked a storm trooper with a 9 mm Schmeiser MP 40 submachine gun, off a balcony one happy day; the man was gut-shot and would have likely died, but he broke his neck when he landed on cobblestones! Remind me to show you a ballistics chart; the impact energy difference is dramatic!"

They stopped at a McD's to use the toilet and switch drivers. Grant bought two orange juices and they went on their way.

"Tell me about fencing!" Amanda asked.

"My high school was too small and underfunded to offer fencing as a sport when I was there. They had had a fencing team but it was disbanded long before we came along. We had access to some weapons that the team had used: dusty, rusty blades, jackets and two masks," he said with a smile.

"Three of us guys, non-jocks, tried to teach ourselves something about the sport. We found a couple books and a bunch of stuff on-line. Fencing is well documented in history before gun-powder; the sons of rich men could afford an instructor. Fencing in that era wasn't any kind of 'sport': it was a survival tactic! If you were skilled in fencing, you could protect yourself against all kinds of lawless... behavior. Only the wealthy learned enough to be effective. A bow was better because it increased the *range* at which you could kill or wound someone, but archery is harder to learn in some respects. Some kind of sword gave you better odds of surviving an attack."

"Anyway, we floundered around with those old blades after sanding off the rust and oiling them. "We were lucky we didn't seriously hurt any of the three of us! The Italians, the French and the Spanish all had different 'schools' of fencing art/science. Later the Germans had fencing schools, often taught by

French or Italians." He looked over at her. "Am I boring you? Not a lot of people are interested in the obsolete skills required to stay alive in the days before guns."

"No; I've watched a couple of fencing matches at RPI, but I didn't know anyone who could explain what was happening!" Amanda replied.

"Fencing is part art and part science; its both offense and immediate defense. It has its own vocabulary. Even today there are regional names for certain parts of weapons; the Olympic sport is too fast for people to see! The weapon-tips are literally *invisible*: they are too small and they move too fast for your eyes to focus on! We put red rubber buttons or wound electrical tape on the blade tips to make them safer. When you put on the metal mesh mask to protect your eyes, it makes visibility *worse*! You can't *see* your opponent's blade-tip, so you *feel* for where his or her blade is, literally, with your blade! When you *think* you know where that blade is, you attack, usually from a 'parry', which is a defensive move that stops your opponents blade, momentarily. It's very fast... you go from 'parry' to 'thrust' or 'coupe' in time spans measured in milliseconds."

"The Olympic sport is all electronic now; the blade movements are too fast for judges to see, so they rely on microswitches and grounding pads to keep *electronic* track of the blades. The uniforms were white to make blood easier to see. Duels were either fought to 'first blood' or death, depending. There are 4 judges and a 'master' who decide, by vote, who did what and to whom!" he explained.

"A guy with a crossbow could stand off and drive a quarrel through an opponent from maybe 30 or 40 yards away; crossbows were easier to load and crossbowmen were easier to train, but they were pretty slow. An English archer, longbow-man, could do the same at 70, 80 or 100 yards. The historical records are very incomplete or speculative, but some massed arrow flights were able to maim men or horses at 200 yards. There's a bunch of misinformation from history, but an English archer could fire as many as 10 arrows a minute compared to only 2 or 3 from a crossbow."

"The three great battles that established English supremacy in Europe were: *Crecy* in 1346 by Edward III of England with 500 longbows, Edward III again at *Poitiers* in Aquitaine in 1356, where John II of France lost the Oriflamme, the French battle-flag, and *Agincourt* in 1415 under Henry V, where he had something like 7,000 longbows! English bowmen were commoners, not nobility; so they were cheaper to feed and much faster on their feet as they wore lighter armor than French or Scottish nobles, who almost always rode to battle in heavy armor. The English yew longbows were heavy bows to draw and release, but they were a lot cheaper than the heavy armor the French nobles wore."

In *muddy* ground, a dismounted noble was really *handicapped* in battle, as happened at Agincourt. Anyway, these three English victories taught the Western World, at least, that the *sword* was a secondary weapon, a weapon to be used when all the arrows had been shot!"

"And then came the onslaught of *gunpowder*: any peasant could shoot through the plate armor of a noble and kill him! Don't bring a sword or dagger to a gunfight!"

"Now, you *saw* your 9 mm rounds disarm that jerk and stagger him to his knees. I don't know the exact numbers, but your Glock probably delivers something like 300 and some, foot-pounds of energy at the muzzle. A .357 revolver might deliver up to... 600 foot-pounds at the muzzle. A .45 auto will deliver something like 370 foot-pounds, depending on barrel length, bullet type and the powder charge. So, while a 9 mm is smaller and lighter and will surely disable a man; the .357 or .45 *might* knock a man down fairly reliably. I'm not all that up on pistol ballistics, Amanda. Your Dad might be!"

"He talks to some of his friends at the office and at cards. They go on and on about ballistics. I sorta tune it out. The Glock did what I wanted; it put him down and unconscious so he couldn't shoot either of us. I was sure he'd shoot you first!" She shuddered as she said that.

"Yeah, I was up next and all I could do was run or try to dive away! I was very glad you drew your weapon and shot him! Thank you again!" He leaned over and squeezed her hand.

Chapter 9   Southern California

Grant was very relieved when he drove into the Manhattan Beach Residence Inn parking lot, turned off the pickup's ignition and sighed, smiling at Amanda. He was even more relieved when they both had been given rooms Amy had gotten them reservations for!

They decided to put her 'stuff' into her room and his stuff in his; they would be safer out of the bed of the pickup while they started work and then started apartment hunting. They would unpack later; they cleaned up and went to The Kettle for dinner!

"It's only 7 but I'm *very* sick of driving!" Grant sighed as he sipped his first espresso.

"We were in no shape to look at apartments! We can start that tomorrow evening. I want to get into my office and start work! We need to start learning our way around!" Amanda replied.   "I'm tired from watching the road and the other cars!"

"Roger that! You liked going through the mountains, didn't you?" he asked, enjoying the espresso.

"I did; I never saw them up close like that; there wasn't much of a snowcap, was there?" she replied.

"No, I think it's another hot, dry summer... more drought seems to be the new norm for the intermountain west. L. A. County is on water restrictions again," he pointed to the little laminated notice in the menu that asked diners to request a glass of water when they wanted it. Water would only be served on request, not automatically served in this restaurant.

"Utah, well the little part we went through, anyway, looked pretty green. There were a couple of small rivers running pretty good. Is that snowmelt?" she asked.

"I don't know, it probably was. The Green and the Colorado both had plenty of brown water, at least along US 70. I suppose that the agricultural people have learned how to store some of the annual flow in reservoirs so their fields can get some water until they get Fall rains. I don't know a lot about farming; I know it takes a lot of water. Some of them pump from wells, of course. A good, deep well might be all they need; I don't suppose there are too many of those," he mused.

"Where's L. A. get it's water? Do you know?" she asked.

"I read an article about the 'water wars' of the 1920's; a group of L. A. businessmen bought up a bunch of land around the Owens River valley up north, just east of the Sierras. The Owens Valley got a *lot* of snowmelt in the spring. These guys bought up land that wasn't much for farming just to get that snowmelt. Years later, it came out that it was all the property of L. A.! They did a major rework of the river channel and ducted the water down here. There's a system of canals and aqueducts you can go see today; it was a very big, carefully planned operation to *guarantee* water would be available down here. They planned pretty well; L. A. has a lot of people that need that water. They may not have green lawns, but they've enough water for cooking and bathing!" he explained.

They went to bed exhausted from the long drive out from Ohio, very glad to be off the road and ready to start their new jobs. Grant was asleep in minutes after he showered.

The next morning they ate breakfast in the Residence Inn breakfast room: croissants, eggs, espresso and orange juice and then drove to the NG campus. He dropped her off in the 'upper' R6 parking lot at the guard shack; she gave him a quick kiss before heading in the first time.

She smiled and blushed. "I can do better than that later, OK? Call me sometime and tell me how it's going," Amanda asked with a smile. He decided she was nervous about having NG people see her kiss him.

"I will; maybe once I've got a desk and a computer!" Grant replied. And then he drove on over to the parking lot between 01 and R5, where Sali had told him to park. It felt really strange to be doing this the first time! His new badge got him into M3 where he went upstairs for the second time.

Chapter 10   Starting Work

Sali met him coming up the stairs and greeted him: "Hi there! I got you a computer, but the IT guy still has to set it up and check it out. Wait for him! He'll get you started once its up on our intranet. Your phone is active so you can get both regular external and internal calls. Paula's in her office; I'm sure she'll want to talk to you!" Sali finished.

Grant went into the 'side office' off the department bay. His two roommates were not in; one computer was up, displaying a Microsoft screen-saver, the other screen was black.

He sat down at his desk, adjusting the chair, which was pretty comfortable, and put his leather folio beside the 'new' desktop comp, cords still wrapped around it. He started emptying old odds and ends from the previous desk-owner: paperclips, staples, a couple of yellow pencils, a couple of pens, a very old eraser and a set of white-board felt markers.

He half smiled at how ordinary it seemed in the silence of the small room, hearing the sounds coming from the big machine shop below him: the whine of hard metal being machined penetrated the drywall of the office walls easily!

He was surprised to hear a cart being pushed into the office. "You Grant Porter? I'm here to get that machine up and running! It'll take me a few minutes and then I'll show you some of the ropes! "I'm Paul Borden, IT! Welcome to Space Park!"

Paul busied himself following a routine procedure of plugging the computer into 110v power and then connecting to the NG intranet; the intranet cable connection was iffy enough he searched the wall for a better connection socket. When he had the computer up; he spent almost 15 minutes checking various apps to be sure they all worked. Then he had to find a printer connection that worked to the printer behind the door. All three desks used a single printer. At Purdue, the dorms had a printer in about every other room; the residents always had to share access.

"Is there a simple program for 2-D sketching like EasyDraw, I can use?" Grant asked.

Paul nodded. "Let's see if its in here... should be!" It took him several minutes to pull up that basic drawing program. *That* was good to see! Grant wouldn't have to make hand sketches which would be only partially legible; he expected to be making colored sketches very soon! He could see the cluster of Microsoft apps: Word, Excel and Powerpoint. Those he knew he could use at some intermediate level of skill. He'd probably learn more about them here! His desk phone rang!

"Hi! You're in?" Amanda asked.

"I am; in a few minutes I should have a working comp! The IT guy is wringing it out for me! Is yours up?"

"The desktop comp is up, but I don't have a cable connection to a printer yet. Alice has a work order in; I should be able to print later today. There are a lot of offices over here: they're mostly doubles and a few singles; the leads get single offices, usually with windows. M & P is having a birthday for one of the engineers today; I think I should go to that to meet more of the people I'll be working with. I've got an org-chart: there are 34 people in the department counting leads, engineers and techs."

"The Mechanical Lab is a big place: all kinds of ovens, test chambers, testing machines. They have their own metallurgy lab where they pot, saw and polish specimens for optical analysis: that lab has 8 techs! That's going to be very useful! I think you can bring specimens over here, Grant!" Amanda said, enthusiastically.

"Sounds good. I'm going to get called into talk to my boss in a few minutes; enjoy, I'll pick you up at the guard shack whenever you say... 5-ish?"

"OK, see you then!" and Amanda rang off.

Paul stood up, gesturing that Grant could have his chair back. "You should be good to go! Call this number and talk to one of our phone techs if you can't get something to work. If you need a 3-D CAD program, we'll need to install a program for it; you might want to request a 27-inch display in that case: the menus and border tools are pretty small and complex!" Paul waved and pushed his tool cart out into the hall as Sali beckoned him toward Paula's office.

"Welcome aboard, Grant! How did your trip go?" Paula asked.

"Well, we're out here; we need to start finding apartments this evening! We had some unnecessary excitement coming out!" He gave a short description of being attacked by bikers coming through Illinois, just west of Chicago.

Paula listened in silence and frowned; "women are usually easy targets for... jerks! Apparently, your friend is *not*! Big surprise!"

"A very pleasant surprise! I got the first jerk with a piece of pipe, but I didn't have much of a plan for the second, the one who drew a gun! She saved my life, probably!" Grant related.

"I look forward to meeting her! She knows she can't carry in here?" Paula asked with a smile.

"She does... the gun's stashed in a good place. She shouldn't need it in here!" he replied. The Glock was out in his truck, but he wouldn't tell her that. They needed to buy some cleaning solvent, patches and a small cleaning rod and clean it soon!

"Good! Let's talk about your first assignment. I assume you got a tour of M3's Package manufacturing shops: mostly the machining, plating and feedthrough installation areas?" she asked, pointing downstairs.

"Well, Robert McPage walked me through the mechanical shop and we looked at the Plating Shop through the windows; he pointed to the feedthrough shop, but we didn't go in."

"OK, you can do that soon. All our packages start over here; a very few are bought out of house. Manufacturing Engineering plays a part in several manufacturing jobs and shops: the Composites Shop where we build mostly antennae, the Hybrid packaging shops and a... call it a 'supporting role' in bonding assembly. So, sometimes, for some projects, we assist in what some call 'resin engineering': the nuts and bolts of organic resin technology. Some of that is in support of the Composites Shop where developmental structures are designed and fabricated: dish antennae, rocket bodies and components and some aircraft structures. NG doesn't build launch aero-fairings; we build pretty much everything that goes *inside* the fairing. We do a *lot* of support for the Hybrid microelectronic manufacturing lines." Paula paused.

"Jerry James, over in M5 Hybrid Assembly, mentioned that when Robert

McPage walked me through their big lab," Grant replied.

"Good! We had an engineer here some years back, senior guy... very broad knowledge of engineering; he got us out of the *mixing* of epoxies and got us back on track to do the *bonding*, one of our core specialties. We were really bad at *mixing* epoxies; all our epoxies, bonding and coatings are now bought premixed and frozen. You'll need to learn about that list of consumables as you settle in. You know that silicon, gallium arsenide and gallium nitride chips are usually bonded down inside a metallic hybrid package with some kind of conductive epoxy?" Paula asked.

"I do; those were traditionally silver-filled epoxy pastes, I have read," he said, smiling.

"That's correct; we used to mix and apply silver conductive epoxy pastes: a list of several different mixes. What we're going to do, because no one ever got funded for it before, is to make a formal comparison of our *silver-filled* conductive film adhesives with one copper-filled, one gold-filled and a new *diamond-filled* conductive film."

"I want you to design an experiment to compare and *rank* silver, gold, copper and diamond-filled films for function. This task will amount to a *qualification* of the material we've never used before. I want you to interface with Jerry James's people in M5 and with Dr. Morgen's staff in R6 when we do this qualification. A 'qualification' is a formal study, that, if we do it properly, everyone can buy in. OK?" she explained.

"Diamond dust is touted to be 'wonderful' for through-conductivity: getting chip-generated heat out of the hybrid and into the support structure of the satellite, usually. If this diamond-filled resin film, *SuperCon*, is good, we want to *use* it. If it has issues, we want to know what they are and if we can use it effectively. It's expensive; well beyond the cost of gold-filled film. If it's only *marginally* better, we'll make, NG Space Park staff, will make a decision based on cost. If it's significantly *better,* we'll figure out how to introduce it to routine manufacture. So, *you* need to learn how to rank it; you need input from both the Hybrid Assembly staff and the M &P staff to do that. Once we spend the money, I want them to be able to buy off on it and agree to the testing that will let us rank it. Does that make sense?" Paula asked.

"Yes, it does; I'm new at this, but it would seem to call for a standard product we already make and have guidelines for as a *model*; we then make identical... modules with your four conductive epoxies: silver, copper, gold and diamond and test them in exactly the same way to make the comparison. I don't know what parameters we test for. Hybrid Assembly has an electronic test group somewhere, I'm guessing?" Grant asked.

"They do; they're over in M5; they are... just north of the Change-room, toward the

144

cafeteria. There's a whole bay with nothing but electronic test-fixtures inside. The Spacecraft Electronic Engineering Group is scattered around in both M5 and R6. Sali needs to get those buildings put on your badge; you're going to need to get into them to get them to teach you what they want to test for in your comparison of the test article. I think you want to meet and learn from a guy named Arty Williamson, he's a big, tall, skinny EE. He lives and breathes RF module function; he'll know what to test for to learn about whether or not diamonds are an EE's best friend!" and she laughed at the cliche.

"OK; you want me on this *immediately*?" he asked.

"I do; use this *JN* or job number," and she handed him a note with a number written on it. "Sali will show you how to fill out your daily work-log on your computer. We fill that out at the end of the day, usually; Payroll, *accounting* really, cuts you a check based on that work-log. Sometimes, they get all squirrelly about how often you have to fill it out. I think their rules say you have to fill it out several times a day if you have a *list* of different charge numbers. If you get called-in to work on an issue, maybe by Spacecraft Electronic Engineering, maybe by the dish-antenna folks, maybe by some of Dr. Morgan's staff, they'll *have* to give you a JN number before you can actually work it. You need to keep a list of JN's, but keep it in a desk drawer you can lock."

"If you have questions about how to charge your time, ask Sali first and then me. The issue comes up more often in Dr. Morgan's group than mine, but it'll come up! Our relationship is usually smooth; we consult with Howard all the time. You may work some of his issues with your friend Amanda McCormick; they've recently had a number of senior people leaving. Some of their folks, like your friend, are young, newcomers; some of them are just a bit better than beginners. You'll run into that soon enough. We, NG, has to hire enough newbies to operate; we mostly learn OJT from each other! That usually works, but it's sometimes got jagged edges, as we confirm what works best," she said with a smile.

"If you communicate well, and Robert says you do, you can help some of those young people who'll be in your meetings. Our work gets really... complex, technical; you'll end up teaching me things I haven't yet learned and I'll teach you some things. I was an Engineering Science major that drifted into management, but I still like to understand the details."

"As a general rule, Grant, you can't assume our people know *anything*; you have to determine if what they know matches what *you* know. Some of them will teach you." She handed him a thick manila envelope. "Here's some literature from Able Bond; this is sales literature, but they're usually very good with facts about their products. Larry Elgin, one of Howard's staff, who is dotted-line assigned to Arty Williamson, an EE, is the most knowledgable person about bonding materials since some other folks retired. He's a Chem E. You might work up a draft plan and let both me and Larry, or maybe Arty, review it. I know he's got some sample diamond film over in an epoxy freezer in the Hybrid Assembly area. He won't have much, but the local sales guy will be happy to talk to you about it! You might contact him when you have a draft plan; he might just tell you who else in the South Bay is using it and for what!"

Back in his office, Grant sat down, pulled the keyboard to his 'new' computer into place below the display, called up a blank document and entitled it: 'SuperCon 1-11 Thermally Conductive Bonding Films Study'. He thought for a minute and added 'Draft Qualification Comparison'.

He studied the blank page for a moment and added a line below: 'G. D. Porter, Manufacturing Engineering, M3'. He started a new paragraph: 'Objective: compare and contrast qualified silver, copper and gold-filled epoxy bonding films with a new SuperCon, diamond-filled 1-11 Thermally Conductive Bonding Film.'

He added a 'cc' line and put P. N. Hopkins name there; after a moment he added H. Morgen to that line. He called Larry Elgin's number as shown in the electronic phonebook in his new computer.

Larry picked up his phone: "Elgin here!" Grant explained who he was and what he wanted to discuss.

"C'mon over! I've got a 10:30 meeting but we can talk before that. Dr. Morgen warned me you were going to call to talk about SuperCon 1-11! I've got a couple square inches as a sample. We should look at it!"

Grant grabbed his folio, made sure his pen was in his shirt pocket and headed out of the office, stopping to see Sali. "Hey, can we get my badge updated to get me into M5, please?"

Sali led him down to the lobby cubicle that served as a Security Guard location for M3. The Security Guard studied his badge, consulted a list of people on a computer screen and then added M5/R6 to the badge. He printed a new laminated badge for Grant and nodded, all in silence.

Grant thanked Sali and headed out the door, clipping the picture badge to the front of his light blue shirt. He paused long enough to be certain he knew where to go by studying a big campus map on the wall. He walked across the parking lots toward the gas station at the intersection of Marine and Aviation. A few minutes later he reached the door into M5 at the South entrance. The door opened to his new badge!

He went down the first corridor until it dead-ended on a corridor. He went down the corridor and found the long row of windows that allowed him to locate and see into Hybrid Assembly where he'd been before. A few minutes later he had located Larry Elgin's office across the hall from the Assembly area.

Larry stood up at his desk and shook hands with Grant. "So tell me something about your career so far?" Larry might be late thirties or early forties, chunky with red hair, Grant guessed.

Grant obliged with a short summary: he was a newbie materials engineer with a background in metal casting, assigned by Paula H. to do a qualification test on SuperCon 1-11, diamond-filled epoxy film. "I'm thinking we want to build some identical hybrids using 4 film adhesives and see how they rank for certain parameters I don't know how to select!" Grant replied with a smile.

Larry smiled. "OK; we might choose a PDSUb hybrid; that's a largish hybrid with 5 substrates and 5 gallium arsenide chips; we've got plenty of performance data for that package. We might look at a couple of parameters to see how well the new material works; one parameter might be interior temperature under load. Another might be noise figure; another would be ... frequency response and power factor. The PDSUb has a thick enough aluminum wall to let us measure interior temperature fairly accurately; it heats

149

up when we turn it on and holds that temperature in the mass of the package before dissipating it to the spacecraft structure. Let's walk over and go into the test area. We can go in the hall door and skip gowning-up if we go in that way."

Larry led him down the corridor that went to the M5 cafeteria and through a card-lock on a numbered door. Inside there were 4 rows of test apparatus arrayed down the room. Each of the test apparatuses appeared to be different. Larry walked past several test sets until he stood beside a big, 30-ish woman leaning forward on her chair making an adjustment to a spider-like arrangement of sharp-pointed needles pointing down in a circle surrounding a hybrid package with the lid off. She positioned each probe with minute X-Y adjustments to touch the center of a certain metallization pads inside the package as viewed through a stereomicroscope.

"*There*, you little bugger! Now *talk* to me!" she muttered and when a series of traces appeared on a signal generator display, she smiled to herself. "*OK!*" she grunted, apparently pleased with what she saw.

She turned around and nodded to Larry. Larry introduced Grant to Ronny Harmon. "Ronny's one of our better test-techs. Grant's brand new, just starting; he's a materials engineer. He wants help setting up a comparison to look at the differences between different *filled* epoxy film die-bond adhesives.

He's asked us to help him design *how* to make the comparison. I suggested we make a series of PDSUb's, each subgroup with a different adhesive film: production silver-filled, copper-filled, gold-filled and this new diamond-filled film. The best one should let us see lower temps in the package and maybe better noise factor and signal-to-noise ratios. What would you do, Ronny?" Larry asked.

Ronnie sat back down and scanned the readout from her test set and made some changes to the input to her little circle of probes. "Those are gold-plated tungsten probes?" Grant asked.

"Roger that! Good for a newbie to recognize a probe card! Where'd you go to school?" Ronny asked.

"Purdue; I didn't study much EE-stuff, but I recognized the probe card. It's a very unique way to get a bunch of electrical signals into a small area," Grant replied. This woman could do complex electronic tests he'd have trouble understanding.

"Yeah. You might get some answers with the PDSUb but you might consider the small LNA, too. Because it's so small, it's more sensitive to package temperature. The Thermcon disc in the LNA package floor helps dump heat out of the package once it's clamped down to the spacecraft slice. If your diamond film works, it might run cooler and show a lower noise factor. Hell, you could build some of each! Every package is going to react a little differently. The Linear Integrator might be pretty sensitive to chip-generated heat, too. How much money have you got?" Ronny asked.

"I don't know; Paula Hopkins didn't tell me that yet; she did give me a JN. Do you need a JN to give me an email suggestion, maybe with some explanation of specifically what we'd look for in testing? I've started a draft proposal, but I need your input make it more real, please," Grant asked.

"I can diddle-up a test outline for an hour's charge to your JN! The diamond-filled film is expensive, I'm guessing?" Ronny asked. Grant nodded. "Keep the film on the thin side: .001 to .002-inches thick; that might help limit *heat-spreading*. The Hybrid techs can cut the film line-to-line so it minimizes any run-out. When we have temperature sensitive chips, we sometimes ask for a thermally conductive 'skirt' *around* the chip to help *spread* heat. In this case, I'd think you'd minimize the area, a bit, to see the effect on transconductance," Ronny suggested.

"You know, we've seen some *hot spots* under gallium nitride chips. The LNA has a GaN chip; that might help light-up the effect of a better conductor. How conductive is this stuff, Grant?" Larry asked.

Grant opened his folio and pulled out a spec sheet on Able Bond 1-11 diamond-filled epoxy film. "This table shows it at something like 1,100 watts per meter degree K. They don't explain how they tested it to get that number. Their gold-filled film is about 450 watts per meter degree K. That's almost 2.4 times better; I don't know how it will actually work: that's why we need to build it into something we already understand and see what we can measure," Grant replied.

"Yeah, they're going to paint it good to sell it! We need to see if it really delivers 1,100 watts per meter degree K. If it does, it might be *very* good," Larry replied as Ronny scanned the data sheet.

"This says they're developing another diamond filler: they're hoping to see 1,500 watts per meter degree K! Damn, that'd be 3 times the gold-filled stuff! We might want to test *that* sometime," Ronny suggested.

"Didn't M & P test a *copper/diamond composite* heatsink material with a conductivity of... I don't remember... maybe 600 watts per meter degree K?" Larry asked.

"Yeah, I think it was a Sumitomo product; it was very expensive and they were all Nipponese about using it in anything like a weapons system! I remember a white paper that satellites were *not* weapons systems! Some kluck argued that satellites were used to *aim* Predator and Reaper missiles and that *made* them weapons!" Ronny replied with a shrug.

Grant had no idea if he needed to consider the ITAR regulations; he'd better ask Paula about that. Able Bond and SuperCon were made in the US, so maybe the ITAR regs didn't apply? He didn't know. Somewhere, NG would have lawyers and application people who studied that!

"I'll send you an initial draft proposal by email; you can comment on it and send it back! Thank you!" and Grant decided he'd go back to Hybrid Assembly and look over their shoulders while they actually *did* some bonding of film adhesives. He'd never seen a tech or operator perform die or substrate bonding; he needed to know more of how that was done. He got Larry to direct him to one of the bonding operator leads. He gowned-up and stood behind her while she populated a small group of LNAs using .002-inch thick silver-filled film epoxy.

The procedure was simple: she removed an 8-inch by 12-inch sheet of epoxy from a freezer controlled to -40 degrees F. She allowed the sheet to warm to room temperature, about 25 degrees C. The material had the consistency of cheese when warm. She used an Xacto knife under a stereo microscope to cut a piece of adhesive the same size and shape she wanted. She typically did this by placing the substrate on top of the upper release-liner and cutting through the top release-liner and the adhesive. Then she lifted the piece of film off the lower release-liner and transferred it to where she wanted it in the hybrid package; she very carefully placed the shaped adhesive where she wanted it. Then she placed the substrate *on* the adhesive with exacting alignment to ensure no over-hang.

The completed package was placed in a vacuum bag and heated to about 125 degrees C. The adhesive film melted or slumped to wet and make contact with, and adhere to, both package floor and substrate. After a specified temperature 'hold', the epoxy adhesive cured, so the substrate was permanently bonded in place. The next step in population was to connect the chips and substrates to feedthroughs with either gold or aluminum bonding wires. At that point, the package could be electronically tested for function.

She pointed him to a copy of the document she worked by on the computer above her work station. "This a 'FIPP' or fabrication-inspection-process-procedure; it defines how we do the bonding. You should get a copy of it for your own understanding. Ask Larry!" He wrote the specification number down on a piece of yellow lined paper in his folio. He wondered about the procedure; each operation would likely have its own 'FIPP'.

He went out to Larry's office and asked about FIPPs. "You need to get your boss's secretary to get you an NG tablet. You can download all the FIPPs to the tablet so you have them at hand to study; you could carry them around in printed form, but that'd be awkward once you get up to speed on the whole sequence," Larry explained.

Grant watched the bonding application and curing and then the operator, Jin-Sue, let him examine the resulting bond-line which was almost *invisible* under both substrates and chips. It was so *conforming* to the chip or substrate, he almost couldn't tell there was anything there under the chip or substrate! It was only visible at 40 X magnification as a kind of glossy edge! Impressed, he thanked her and went outside to see what the M5 cafeteria offered for lunch.

He decided he didn't need anything more than a McD burger, so he drove down Aviation and pulled into a smallish parking lot beside a small Golden Arches sign. He got a cup of diet cola, a burger and a small fries. He sat in a booth facing east down Manhattan Beach Boulevard, feeling a bit disoriented in the bright California sunshine. He was just starting in a whole new place!

Chapter 11 Designing An Experiment

Grant went back to his office and worked on the document he'd started earlier. He built a Word table showing the PDSUb, the LNA and the Linear Integrator on the left vertical axis and a series of columns across the top horizontal axis. He made the first column 'package temperature' under load; the second column was 'noise figure'; the third column was signal-to-noise ratio, the fourth column was 'adhesive filler material'. The fifth column was 'film thickness'; the sixth column was 'comments'.

He spent some time arranging the table grid and the grid-boxes for size and fit on a landscape document so he could add data whenever he got it. He studied it for appearance. He added a footnote at the bottom for costs. He'd need to get the vendor to tell him that; he'd better see if he could reach the salesman whose business card was in the brochure Paula had given him.

He called out on his new office phone, reaching a telephone receptionist. "Give me your name and number and I'll have William Manchester call you back. He's in the field; be patient with him. He will try to call you just as soon as he can!" Grant gave her his name and his new number.

Grant sat back and tried to decide how many of each package they should build. He had no idea what the material would cost, so it was hard to decide how many they might make. He decided that 9 of each of the 3 devices for a total of 27 test packages, might be a place to start. Paula might object to that number if the material was really pricey; he might come down to 7 each, would be only 21, if she did.

He'd taken a course in design of experiments at Purdue. He should dig out that textbook and bring it in to be able to set some boundaries. He thought Sali locked the room every night to prevent other employees from entering the office. He might bring most of his textbooks in and keep them here; he wondered about that. He knew he needed the Handbook of Chemistry and Physics. He might want his text on Ceramics and the one on non-ferrous metals, which was now more important than the Ferrous Metallurgy course he'd waded through! He might want the Metals Handbook and the Materials Science and Engineering text, for starters.

A *real* engineer had references and detailed specific data ready at hand. He'd better get some specific data on the 4 materials in his experiment! He had a text on injection molding, but that text wouldn't have anything about epoxy bonding films, especially custom-formulated, thermally-conductive films. He'd have to rely on this William Manchester for the data the vendor provided. He wondered if this guy was as technically competent as he thought his Dad was at technical sales.

That thought made him recall the telephone conversation the two of them had had from the motel in St. George, Utah on their way down into Southern California. His Dad had been impressed Amanda had been able to stop the second biker in his tracks! He had invited her to hunt with Grant and him sometime when that might be arranged!

His phone rang! "I'm William Manchester from Able Bond! How can I help you, Grant?"

"I'm new at Northrop Grumman; I've been tasked with comparing Able Bond 1-11 to some of our standard conductive epoxy bonding films. I'll need some of your data and pricing. Maybe we should plan to meet here sometime soon. Might we do that?" Grant asked.

"Yes! I can meet you tomorrow afternoon at 2 PM; I've got a luncheon commitment, but it's a quick one. Would that work?" Manchester asked.

"Sure; let's meet in the lobby of M3. Will that work?"

"I'll see you then! Thank you for calling me, Grant!" and he rang off.

He asked Sali about a room key and when the room should be locked, explaining he wanted to bring some books in to have them readily available.

"You can generally leave books out on your desk safely, Grant. Not a lot of people come up here, but both John Lahti and Leann Davies, who's on vacation, your roomies, have keys. I had a new key made for you!" She handed him a bright and shiny brass key. "I generally unlock it when I get in and lock it when I leave," Sali explained. He slid it onto his truck keyring.

She told him who was located in the next office bay, where some more of Paula's people worked. He made a copy of his draft plan and asked her to show it to Paula. He had highlighted and bolded the word 'draft' so she didn't think he was nuts!

He only had the one heavy box of text and reference books in the room at the Residence Inn; he had already marked each of them with his monogram in silver ink on the book spine back at Purdue. He'd check the box and decide which books to *not* bring in.

He was surprised when Amanda knocked on the door and walked in, smiling! "Hi there! Are you ready to eat or do you want to go out to look at apartments before we eat?" she asked, catching his hand but stopping short of kissing him.

"Hello! Sali, this is Amanda McCormick, my... friend. She started today over in M & P!"

Sali's face was blank for a moment. "*You're* the one they're calling *New Girl*! OK, I got it. I didn't realize you had a... woman friend, Grant! So you're going to have a pipeline directly into M & P. That might be OK! Good to meet you, Amanda; welcome to Space Park!" Sali replied, eyeing Amanda carefully.

It was just 5:15 PM. "Hey, is Paula going to want to *talk* about the draft I gave you? Should I hang around for her to come back?" he asked.

Sali shook her head. "She's in a meeting in R5; it'll probably run until 5:30 or 6:00. I'd leave and go find a place to live! She can check-in with you tomorrow!" Sali replied.

"Tell her I'm meeting the Able Bond salesman tomorrow after lunch; I'd think she'd want to know that. See you in the morning!" And he closed and locked his office door, waving good evening to Sali.

They started down the stairs to walk out to the truck. He whispered: "from the look you got, she's jealous already and your peers are talking about you! You're impressive enough they're upset! I suppose that's par for the course?" he asked, not sure he wanted to hear the answer.

Amanda made a face. "I'm a newbie engineer; I don't care what they think of me as long as they accept me as a competent engineer," she said in a low voice that wouldn't carry very far.

"'New Girl', huh?" he teased, hoping it was OK.

"Yeah, some of that is expected," she said with a quick grimace. "There are more women over in Morgan's department than I

161

expected. I'd think they'd be more interested in ability than looks! We'll see! So what should we do? Drive on out to those two places you looked at? We could eat out there, I suppose?" she suggested.

Grant headed out on the 91 Freeway in the bumper to bumper 'going home' traffic. A lot of people couldn't afford to live near the Coast; they commuted West in the morning and East in the evening.

It took just over an hour to reach Tesseract Towers in Yorba Linda. They parked in the lot and walked into the sales lobby. Grant gestured that Amanda should ask questions of the young Asian sales agent, a different person than the one he'd talked to earlier. The young woman showed them the three model apartments behind the small lobby of the first tower, and answered Amanda's questions about the rent or lease agreements.

Grant liked what he saw; he'd maybe go for the smallest furnished unit to save money. He didn't think he do much cooking-in. He'd do his reading in the small living room off the kitchen in the evening; he didn't watch much TV anyway. He'd likely eat breakfast in: cereal, toast, orange juice and espresso. He had no idea of how Amanda would live in her apartment.

Amanda wanted to go up in one of the buildings and look at an actual rental 'studio', not the sales units, that had never been lived in.

The saleswoman got keys and they went up in an elevator to the 7th floor where 2 studios were available. "You have a generator for when the power's out?" Amanda asked.

The young woman nodded: "our generators, one for each tower, are automatic. They come on in less than a minute. If you're in an elevator in a shake, it comes down to the lobby, pauses before resuming service. It's all automatic!" she said.

Grant had no experience with earthquakes and was not looking forward to the first one! There were two stairwells in each tower; so one *could* walk up and down. He decided that he didn't need to live up on the more expensive two upper floors where there were fewer units so they could be a little larger! That would be a longish walk with the elevator out!

The sales agent just realized that they wanted *two*, separate, studios; they were not living together. Both of the units she showed them looked new, barely lived in; one unit had just been repainted in pastels: light blue and beige; he could smell the latex paint.
The unit was bright and airy with big picture windows. Grant liked the bright appearance, but he didn't think he'd spend a lot of time in an apartment. Work days he'd be in Redondo Beach; play days he might be out ramming around Southern California, seeing things he couldn't see back in Ohio or Indiana!

Amanda opened drawers, cabinets and the wardrobe, visualizing living here. When she was finished she got a copy of the lease schedule and a brochure. "OK; I've seen enough for now. Thank you very much, Ms. Clark! We'll be in touch!" and the two of them headed outside.

"There's an Outback; how does that sound," Grant asked. Amanda nodded acceptance of that and they walked across the parking lot to the Aussie-style steakhouse.

They got a booth and ordered drinks; he ordered espresso and she ordered ice tea. "Those apartments are small at only 750 square feet, but they're new-looking and nicely designed. I think I can start in such a place. I never really considered how much space I need. I'll probably only eat breakfast and lunch at 'home'; dinners I might just eat with you! Are you all right with that?" she asked in a low voice so no one else could hear, smiling.

He smiled and leaned towards her across the table. "I could be *very* good with that, Amanda. Are we an 'item'?" he asked.

She grabbed his hand and smiled as she said: "yes, I suppose we are! I think we might work on that some... maybe even later!" She blushed. "I don't know why it's hard to talk about that. I want this to grow, naturally, whatever that means! Don't you?"

"I do; I looked forward to meeting you for dinner and apartment hunting very *much*," he replied.

They got the drinks then and ordered dinner; she had him go ahead as she perused the menu. "I'll have the small filet, medium with a baked potato and butter and a small green salad," he decided.

"I'll have the same but I want fries, please," she requested. As soon as the waitress left the table he reached for her hand.

"This feels very... comfortable, having dinner at the end of the day together! Maybe we're celebrating new jobs?" she suggested. "Aren't the New Rameno Village apartments similar to the Towers, Grant?" she asked.

"Yeah, very similar; the layout in the ones I looked at are within a few square feet of the ones we just looked at," he replied. "I had an errant thought!"

"What: go together and rent one of the upper ones!" she asked, eyes sparkling.

"Maybe not now, but maybe later," he replied. We need some more time together!" he said with a chuckle.

"We do! We can work on that. I like what I see at work. I had a meeting with Dr. Morgen and 3 of his engineers. One is an older man who calls himself a metallurgist; one of the women is a mechanical engineer and the other one is an EE. Dr. Morgen wanted them to meet me and to get us talking about substrates: alumina, beryllia, silicon carbide and the newer artificial diamond. You ever study SiC? It's not a traditional ceramic at all, it's actually a polycrystalline layered deposit that can take 3 polytype structures; two are hexagonal and one is cubic. It has better thermal conductivity than beryllia, something like 360 watts per meter degree K; beryllia tops out about 330. The chip foundry across from R6 in D1 can grow it under some still experimental conditions," Amanda said, looking off in space for a moment.

"It's not an easy material to grow and it's difficult to cut, but Dr. Morgen wants us to consider testing it because of the higher thermal conductivity. It has a thermal expansion that's a bit low at 4 ppm per degree C. BeO has a TCE that's higher, about 8.3 ppm per degree C," she related. "I think the best material is diamond, but it's a lot more expensive!"

"Dr. Morgen's group did some testing of a Japanese composite material: diamond dust dispersed in a matrix of copper. It had a thermal conductivity of something like 650 watts per meter degree K; it worked quite well but it was expensive and difficult to get because of ITAR issues. But anyway, he brought up the study your boss has *you* starting; he expects that *your* diamond-filled epoxy might be a kind of magic bullet for GaN devices. They put out a *lot* of heat! It's very nice to be included in developmental work, isn't it?" Amanda asked.

"Yes, it is! I know that crystalline SiC is called *moissanite* by jewelers. You can put it in a wax ring and investment cast it directly in place, in both silver and gold! I haven't had an opportunity to try that, but I'd like to! It's 9.5 on the Mohs scale, so it's almost as hard as diamond!" he explained.

They enjoyed the steak, which was still pink in the center and quite moist. Afterwards, it was just 7:30 PM, so they decided to go look at New Rameno Village's apartments.

When they found the place, they took the time to walk around in two different 'studios' in one of the four buildings. One of the two apartments had never been lived in. It faced the other 3 buildings across a courtyard that served as a grassy playground. New Rameno had two buildings set up with double bedrooms for couples with children; the other two buildings were strictly single bedroom units. Amanda poked around opening draws and closets as she had at Tesseract Towers.

Grant decided he could live in either place, but he liked the view from TT looking north at the Chino Hills golf course better: all green with a few trees. When Amanda had seen all she wanted to, it was coming up on 9 PM. They headed back down the 91 Freeway towards Redondo Beach. "They're pretty similar; the view to the North from the 7th floor was pretty nice; do you like the Towers?" Amanda asked.

"I do; should we come back tomorrow and sign-up?" Grant asked.

Amanda nodded yes after a minute's consideration. "OK; let's do that, but let's look at as many units as they have. If might be good to be near each other, so we don't have to walk so far!" she suggested.

Back at the Residence Inn, she grabbed his hand and tugged him towards her mini-suite. Once inside, she asked if he wanted anything to drink from the minibar.

"I don't generally drink before bed, Amanda. Some ice water would be fine," he allowed. She got them each a highball glass of water over ice cubes and they sat down on the small sofa together.

He enjoyed watching her. "What?" she asked.

"You're interesting to watch! I hope that's OK?" he asked, not quite teasing her.

"Its *fine*, especially in here where no one can see us!" And she leaned over and kissed him on the lips. "I owe you from that night back in Illinois; I was... a little shocked by what happened. I needed time to process it, Grant. I am... somehow very comfortable with you; that's a good thing. We may get to work together on different tasks; I'd like that. I don't know how to do this, but I like being with you a lot. You... make my day! Are you OK with us sort of... not going overboard right away?" she asked.

"I'm OK, but I'd like to... work toward something closer. You OK with *that*?" he asked, smiling. She leaned towards him and studied his face for a moment in silence and then pulled him into an embrace! The kiss took a while, surprising him a little.

When she ended the kiss, she sighed and just looked at him, smiling. "You OK with that?" she mimicked him!

"I'm very good with that... with you; I hope you can sense that. I'd say it some other way, but I need some time to figure out how to express myself! This is new thing for me...".

She caught his hands and smiled. "Me too; I wasn't a complete loner, but I never met any guy I wanted to be with. There were a lot of jocks and beer-drinkers that I *didn't* want to be anywhere near! You said you don't drink beer?"

He shook his head, *no*. "Maybe I might have one of the hard ciders... once in a while. Cider has a nice crisp taste, to me; it's not strong enough to get much of a buzz from one bottle. I... I'm not comfortable... buzzed, actually; I guess it's just a bit irresponsible! I hope that doesn't sound too... strange?" he asked as he studied her very pretty face.

"It sounds fine. I like to be in control, too. I'm not comfortable being buzzed either!" she replied, nodding. They spent some time on small talk, holding hands and sharing a kiss from time to time.

He could feel that she wasn't very eager to kiss them into sex; he didn't want to get *that* started, because he wouldn't know how to *stop* it! It was very nice to talk and 'get acquainted', whatever that meant between two young people in this day and age. He guessed that neither of them was very casual about anything, sex included.

He decided he needed to get some sleep so he could function tomorrow. "Amanda, let's get some sleep; we can spend some time together again tomorrow evening. OK?"

"OK, Grant; I suppose that's the reasonable thing to do!" They got up and at the door, they embraced... an awkward 'good night' thing without any real heat.

He went to his room and took his usual shower before going to bed. He lay there, not more than 30 or 40 yards from her, wondering what it would be like to get carried away and have sex with her! He wanted that, but not *now*; he wasn't ready to abandon good sense and throw himself into that kind of relationship. He didn't know her well enough, and she didn't know him. Later would probably be better!

Damn! His mind swam with fantasies about sex with her, but he didn't have any real clue to how to go about it without risking her friendship and esteem!

The next morning, as they met to get into his truck, they decided to try the main NG cafeteria for breakfast, to see how that might work. He parked in the M3 parking lot and they walked across to S-building; he didn't feel any of the tension of the evening before.

He ordered a 'skillet scramble' and she ordered a small omelet. The food was good without being at all fancy; eating here, they could both get into their offices right at 8 AM. He said 'goodby for now' and she headed across Space Park for the Manhattan Beach campus and her office; he headed back to M3 to his office.

Paula waved to him from her desk as he started into his office. He detoured towards Sali, sitting behind her desk; she waved him on into Paula's office. "You have a plan for how you want to do this, Grant?" she asked, pointing to his printed email. Sali would have printed it out so Paula could keep a hard copy for reference.

"Yes, Ma'am; I'm trying to keep it simple: make 9 LNAs with each of the four different filled epoxies and see how they compare for cost and test for performance; that's 36 test articles. The silver-filled epoxy should pretty much exactly replicate the performance we get on routine production; that should be our baseline or reference performance. The gold-filled epoxy should be a little better; I don't know how to estimate how the copper-filled epoxy will group. The *diamond-filled* epoxy should outperform, transfer more heat, than *either* the silver or gold-filled epoxy. "I'm waiting for Williamson and Harmon to specify *exactly* how they want to test for heat transfer, maybe along with some of Dr. Morgen's people, Ma'am."

"'Paula', please, let's call me 'Paula'; I'm not old enough to be 'ma'am'! OK?" she requested with a smile.

"OK, *Paula*! I should learn how to place an order with William Manchester; I'm going to meet with him after lunch today. Jerry James and Larry Elgin assure me they have the 3 cheaper materials on hand in several thicknesses; all we need do is order the diamond-filled material in our choice of thickness and set aside 4 groups of the LNA packages for a... *parallel* build: same package, same operators and same circuitry. Jerry will help me write the build orders correctly," Grant explained.

"Ronny Harmon indicated *she* thought we'd see very different results; I'll hope she's right, but we'll see how the numbers come in. I may learn more from Manchester when I meet with him. Can Sali help me with ordering the diamond-filled epoxy?" he asked.

"Of course she can; I think that the buy order has to come through M5's Hybrid shop. Jerry should be able to help both of you with that. Be sure to explain to the Hybrid people that you're running a 4-way comparison but that you want zero differences except for the epoxy itself! Once you start the assembly, I want a brief daily update, OK?" Paula asked.

"Roger that, Paula; can I do it with an email so we have a written record?" he asked.

"Yes, that'd be good practice anyway; word-of-mouth gets forgotten after a couple weeks!" she replied. "How many labor hours are there in an LNA, anyway?" she asked.

"I'm not sure; something like a morning with some quick electrical testing: 4 or 5 hours. I'll get a hard figure on that. The brochure says the diamond-filled film adhesive cures like both the gold and silver-filled films, so we shouldn't see much offset on labor, if I understand that. I want Manchester to confirm that. OK?" he replied.

"Go to it, Grant; keep me in the loop! We'll meet with Dr. Morgen when we have the first data; I want his buy-in on the preparation as much as on the results! See you!" Paula waved him away.

Chapter 12   Low Noise Amplifier

Grant went downstairs to talk to Pete Franklin about how build orders were written for new packages. Sali had suggested he go find and meet Pete to learn more about those details.

It took him almost 15 minutes to find Pete; he ran him down out on the machining floor where Sumitomo aluminum-silicon composite material was being machined on a turning center. The machine operator had 24 small blank blocks of the low-expansion composite 'metal' on the X-Y table of his big milling machine in a gang-vise.

Pete was asking the machine operator something about how long the 24 packages would take to be machined with multiple cutting tools. The answer appeared on the display screen of the turning center: 48 minutes, based on previous identical runs of making something called a 'Linear Amplifier'. The LNA was a similar, relatively simple design with 3 cavities, and 6 feedthroughs.

Grant introduced himself and asked for 5 minutes with Pete to discuss a new build. Pete led him on into the package assembly lab and they sat down at Pete's desk. Grant explained what they were going to do, emphasizing that 4 different 'flavors' of LNA, 9 each, were to be built with as few differences as possible. All 36 starting LNA packages would be identical; the only differences would be the substrate and chip-attach thermally conductive films applied in M5 Hybrid Assembly. Pete made some notes on his iPad tablet.

"You have to get Sali to get you one of these! It saves paper and I can keep a *lot* of details where I can pull them up quickly when I need them!" Pete exclaimed.

"Yeah, it's on order; I might have it tomorrow. You can do it with pencil notes on paper, but it's less legible and you have to keep track of the paper! Can I *look* at an actual LNA package, please?" Grant asked.

Pete got up and beckoned him to follow; Pete scanned a stainless steel roller cart and pointed to a tray of reflective 'aluminum' blocks. "Get gloves from over there," he pointed. 'You can examine them on that stereomicroscope but put them back where they were, please. This tray will go across the shop to Plating, where they get cleaned, zincated, electroless nickel plated and finally Type III gold plated. Then we install 2-kinds of feedthroughs in them in a single fluxless hard solder process; we install the thermal pad in the central cavity in the same process. The thermal pad is a disc of copper/tungsten generally called 'Thermcon'," he explained.

"Thermcon is a complicated material to work with: it's difficult to machine and it breaks-out, crumbles, unless it's machined in shallow passes. Once we have it soldered into the aluminum-silicon material, we can finish-machine it a little easier," he explained. Pete studied him to be sure Grant followed his explanation.

"The gold/tin solder melts at 278 degrees C, so the thermal excursion to reflow gold/tin also tests the plating *integrity*. We very *rarely* see any kind of plating failure like blistering," Pete explained. "So, this new thermally conductive material should give us an improved hybrid?"

"That's what we think the result should be; the SuperCon adhesive will help conduct heat into and through the thermal pad and out the floor of the package. It'll cost a little more but work a little better. That might be a good trade-off!" Grant replied.

"So, looking ahead: are we liable to see diamond substrates sometime?" Pete asked.

"Yes, I suppose we will; actually, my friend over in M & P is already talking about that. Diamond is expensive, but it conducts heat *very* well!" Grant replied.

Grant spent about 15 minutes examining the unplated LNA package using a stereomicroscope, so he could focus in on the different surfaces. The solder-joint around the thermal insert was 'interesting': the milling tool had cut the aluminum/silicon composite smoothly, leaving an almost-polished surface; the tool had left a rougher surface on the copper/tungsten thermal pad. He studied the interface between the two materials closely: how elegantly simple the design was! You wanted to transfer chip-generated heat out of a metal box? Just install a high-conductivity 'conduit' or pad in the metal!

He started to walk over to M5 to talk to Jerry; he needed to get Paula an answer to the standard build-time for an LNA. In M5, as he gowned up to go find Jerry in the Assembly area, he got a call from Sali.

"Hey, your new iPad is here; I've got it in the right hand shallow drawer of your desk. The desk locks, so it's secure in your desk. Do you want some help in learning how to use it?" Sali asked.

"Uh, yes, of course; I want to know how to download FIPPs so I can have them with me. I'll come back as soon as I get some instruction from Jerry James... maybe 30 minutes, OK?" he asked.

"OK; I'll be here!" she replied.

Grant got Jerry to explain an actual build order for LNAs. Different operators achieved different assembly times on LNAs: the mean time was 3 hours and 48 minutes. The better operators could do the assembly in 3 hours and 30 minutes, so there was a *range* of build-times! Grant hadn't anticipated that.

"When Ronny starts testing *variations* on the LNA, the test times will vary some, too," Jerry warned him. "You can check with her, but I'm pretty sure that a good LNA takes about 20 minutes in test. If the chips all group around a good mean value, they only take about 20 minutes. If she has to tune them individually, adding inductance with ribbons, they take longer. She uses pieces of gold ribbons, .003-inches wide by .0005-inches thick to tune them, but does a temporary tuning 'bridge' and writes down the length of gold ribbon needed and where she wants it. Then she returns them to Annie Nguyen for permanent tuning ribbon bonds; Annie takes about 4 minutes per bond. So, there's no such thing as *identical* LNAs in testing; each one may be slightly different to get them all within specs. They have a *range* of labor content! OK?" Jerry asked if Grant understood what he was teaching.

"Thank you; Sali got me an iPad, so I should be able to start keeping legible, meticulous notes!" Grant walked back from M5 to M3 and got Sali to help him learn how to download the LNA procedure to his tablet, in all it's details. He went to his desk and sat down when Leann Davies, came in and introduced herself. He stood up and introduced himself; he now *knew* one of his 'roomies', at least enough to recognize her. Leann was medium height, slim, with black hair; she looked like a runner. He wondered what her areas of expertise were?

He hadn't expected the hybrid modules to have a range of build and 'tune' times. It was logical, but he didn't yet know *squat* about microelectronics!Hybrid assembly was complex enough it had mechanical, electronic, chemical and microelectronic aspects, *all* of which had to be mastered; he had a *lot* to learn!

Grant went to lunch in the S 1 Cafeteria a few minutes early so he could be ready whenever the guard desk called to tell him William Manchester was in the lobby. Sali suggested he use the small office that served as a conference room just off the front corridor. It had a table that would seat 4 people easily and six with some minor crowding. It had a white board and a printer. He knew some engineers met with vendors in either of the cafeterias because he'd seen them doing that. Some engineers arranged to go out of house to one of the local restaurants for better food and drinks the vendor would pay for. He wasn't ready to do that, as junior as he was!

Sali called him: "Your vendor is in the lobby. Grant went downstairs and met him. William Manchester showed up at 1:30 PM; they shook hands and sat down.

"I'm a graduate materials engineer from Perdue. I know some things and am totally ignorant of others, so teach me! I've done some hand layup of epoxy/graphite and epoxy/S-glass. I've never bonded a substrate down with any of your thermally conductive films, of course."

"Well, I'm a tech salesman; I don't typically do substrate or chip bonding either, but I've watched it done in several of the plants out here. I have a fairly thorough understanding of what's required to make it work properly," he replied with a smile.

"Your peers here at NG went to vacuum-bagging about 10 years ago; vacuum bagging get's you 14.7 pounds per square inch pressure over the whole face of any chip, substrate or package so there are no air bubbles to trap heat. The flat backside of the *vacuum bag* has a Nichrome resistance heater embedded in an aluminum plate, so the operator can set cure temperature to within about .2 of a degree C. The rubber bag forces the assembly against the heated plate, or platen, with reproducible and very uniform pressure; this means we don't see failure-to-wet or air trapped in the epoxy joint. The pressure/temperature uniformity and reproducibility are *very* good! Now tell me how you're going to compare the SuperCon to your traditional film epoxies so I can maybe help you do it," Manchester asked.

Grant explained his simple plan.

"OK, that sounds good; its simple; simple is almost always good! Let me do some simple salesmanship: film epoxies are premixed, cast into sheets between release liners. So, they are *very* uniform in thickness; there's no *run-out* because the epoxy... grabs, wets, *both* the package floor *and* the substrate bottom. The operator cuts the *exact* shape she wants, usually by cutting around the edge of the substrate itself, or a template. So she saves time and material. Because the film is

180

pre-conditioned to *not* flow very far, there's almost zero runout. That is to say, the material is captured on one face by the interior of your package, which is almost always Type III gold. Gold doesn't oxidize, so there's no aluminum oxide barrier to overcome. On the other side is the substrate metallized layer, which is also gold-plated, so it doesn't oxidize either." The man watched Grant's face as he explained.

"I'd call it *idealized* bonding, but that sounds too much like sales hyperbole. It's *nearly* perfect bonding; we can argue about what 'perfect' means! To your operators, this will be a simple study: they already know how to use the material. I expect you will measure something like *3* times the thermal conductivity of silver-filled film adhesive. Perhaps you might share what you actually measure," Manchester asked.

Grant didn't know if he *could* share any data; 'proprietary' meant he probably couldn't share data, even with the vendor!

"I'm new here; let me see if I'm allowed to do that. If my boss says I can share some of what we learn, I'll do it! It's going to take us a few days to get the comparison started; be patient with me until I learn more of how this goes. I do need to know the pricing for your SuperCon so we can order some; can you just give me a price sheet?" Grant asked.

"Here's a price sheet for NG; you're in a group of corporations that share the same quantity breaks. Your purchasing group has a specialist who buys all the adhesives: Marlene Eskivar. Here's her phone number!" Manchester offered with a smile, and a sticky note.

"I need to talk to her so she knows what I'm trying to do! Do you know where her office is?" Grant asked.

"She sits in R1, the old Laser building. I'm sure the receptionist can direct you to her desk," Manchester replied.

"So, we bring the cast adhesives in, stored in foam boxes with dry ice and then place them in freezers controlled to -40 F. We bring each adhesive sheet out of the freezer, let it thaw to room temperature, cut it to shape and apply it. All the adhesives are similar in that procedure?" Grant asked.

"In general, yes; your operators see discrete differences between them: some thicker films tend to *tear* if they aren't cut cleanly. I don't expect SuperCon to give them any issues. Each sheet is marked with a bright red stripe so that our most conductive material stands out from the others," Manchester replied, nodding. "I believe your operators are among the most experienced and skilled in the aerospace engineering... niche; I don't think they'll have any issues. If you have any questions, please call me."

"I'm inclined to order .002-inch thick, 2 mil, material. It might keep the thermal barrier to a minimum; is that logical?" Grant asked.

Manchester nodded. "The thickness is very tightly controlled. 2 mils is going to work just fine! I wouldn't pay extra for the 3 or 4 mil material. I wouldn't use the 1 mil stuff, just In case your package floor isn't completely smooth and flat. *If* you had a machining ridge on the floor, your substrate or chip, might 'bottom-out' on the ridge and form a stress-riser that might lead to a chip-crack," he explained.

Grant nodded; he hadn't checked to see what the flatness/surface finish specs on the LNA *were*. He'd better do that next! He stood up to end the meeting and accepted Manchester's card. "I'll have cards sometime soon; they're being printed. We'll meet to discuss progress. Thank you for meeting with me. I'll call if I have any questions."

As soon as Manchester left, he headed out to the machine shop to find Pete. He'd know what the package floor finish spec was! He couldn't locate Pete, so he went out on the floor to speak to the tech to whom Pete had spoken to yesterday. Harry Bond shook hands with Grant and explained that the standard finish for package floors in aluminum-silicon composite metal were typically 'better than 10 microinches.'

That wasn't all that good, really. Grant raised his eyebrows. "Our gallium arsenide chips are on the order of 4 mils thick... a 1 mil ridge from a cutting insert *might* be an issue.

Do you know what the actual *mean roughness* of the floor is?" he asked.

"Uh, well... we examine them optically, prior to plating, as they come off the turning center; we've never measured them. I'd guess they might go 2 rms mean on a profilometer. Are you looking for a *polish*?" Harry asked.

"I'm new to this, Harry; I don't know, yet, what's 'good enough'! If I wanted something like *1* rms, what would we have to do and what would it cost?" Grant asked.

"Ah, that's smoother than what we get down in a cavity; how far away from the wall can I start my profilometer trace?" Harry asked in reply.

Grant smiled. A profilometer was a linear tool that dragged a stylus across a surface and mechanically magnified the roughness and printed it out on graph paper to show the series of peaks and valleys of the surface finish. Grant had taken a lab at Purdue where he had to learn to run a profilometer on a series of surface finishes to characterize the roughness. 1 RMS was a pretty good polish, optically reflective; good polishes required extra labor and cost more than rougher 'as-machined' surfaces. He didn't know exactly how respond to the machinist. He bought some time with a smile while he considered.

"Is that not smooth enough? No one ever complained about 2 to 3 RMS before. We might hand *stone* them, but we're not going to get into the .030-inch corner radii with the alumina stones we have. I suppose we could go back and do a... say .0005-inch deep crosscut. That'd go fast... maybe 3 or 4 minutes per cavity... 3 cavities or 9 to 12 minutes. I don't suppose *anyone* will like that added labor," Harry replied with a grimace.

Pete walked up! "You got a problem with my machinist, Grant?" he asked.

"Of course not, Pete! *But* I just learned something about surface finishes 'reaching through' the epoxy adhesive film, if that's the way to say it. A 1 mil machining ridge might serve as a *stress-riser* to break either a substrate or a chip in thermal cycling. None of us want to see that!" Grant responded.

"I think we parley with Don McIntyre and maybe Leann Davies, Grant. I can't authorize a *polishing* requirement," Pete declared.

"I surely can't, either," Grant replied with a grin. "I also can't allow any of these 36 LNAs to suffer stress-cracks in either substrates or chips! Let's parley, before I explain to Paula we can't proceed as is."

"You don't need to threaten me with her name, Grant!" Pete said with a frown. He was being territorial: Grant didn't have any chops to question machining of hybrid packages!

"Pete, I don't threaten, much; we're looking to *qualify* a new, high-price, high-conductivity film adhesive. We have to get it right the first time! We may be able to justify a smoother package floor to do that. Let's get it on the table and see how we should proceed.

185

I'll ask Sali to get us 30 minutes or so; maybe we invite Dr. Morgen's M & P group to join us, so they know what we're considering. Do you want to invite your boss, Dan McIntyre, to be present? It might save time to have everybody present at one meeting so we don't have to do it again!" Grant suggested, reassured that he knew most of who needed to be present; he didn't want to stomp around in Pete's sandbox, but he needed to establish that he could reason and step towards the *real* requirements of this new product, whatever they turned out to be!

"I don't want to give anyone the impression we don't know how to machine LNAs; we've been doing this for years!" Pete replied with a frown.

"No one's suggested that; we may have a *new* requirement for LNA's, that's all. Let's parley!"

Sali managed to get them a meeting at 4:30 PM in a small conference room across the hall from Paula's office. Paula introduced Grant to Dr. Morgen and Don McIntyre, neither of whom he'd met before. "This is Grant's meeting: Grant, spin us up!" Paula asked.

Grant introduced the meeting with a simple table projected on the white board from a transparency Sali had made him, of how the 4 groups of LNA's would differ *only* in the conductive epoxy film. He didn't know these people: this might be complicated!

"For discussion purposes, we want to use the thinnest and therefore least expensive form of diamond-filled sheet epoxy. We expect to be able to dump *more* chip-generated heat through the thermal pad in the LNA floor and out into the external heat sink, so the LNA runs cooler. If we can *achieve* that, we might see higher performance; we expect we might achieve a lower noise figure and maybe an increased signal-to-noise ratio. Any questions about that?" he asked. No one had a question.

"OK, an issue, William Manchester, the Able Bond sales engineer, brought it up, is that to use a *thinner* adhesive film saves cost, but it may allow machining ridges in the package floor to *reach-through* the epoxy and let the chip or substrate touch the top of such a ridge. Such a ridge, maybe .001-inches or 1 mil high, *might* form a stress-riser that would *crack* the chip or substrate, maybe during thermal cycling. We want to prevent that! Pete and Harry, one of the machinists; suggest that the specification for surface finish of the package floor is typically 'better than 10 rms, root mean square'. So, that tells us that we don't *know* what the surface finish actually achieved is. It's in the range of 2 to 4 RMS. Have I said that properly, Pete?" Grant asked.

"Yes; because it isn't *specified*, we don't do any measurement or secondary finishing. Secondary finishing will take more time and add to our labor cost. We *could* make a final, finishing pass with a ball-end mill and get a smoother finish... maybe a .0002-inch finish cut, which might reach about a 1 RMS polish, but that would add some 3 minutes per cavity. In the case of the LNA with 3 cavities, that would add about 12 minutes of labor," Pete replied.

"OK; none of us wants to *add* labor to our LNA cost. However, if we achieve *better* electronic performance, that added labor might be a good trade! Pete suggested that the cavities might be hand *stoned* to reduce any ridges. Stoning is done by hand with, in this case, the smooth end of a ceramic, usually alumina or silicon carbide, cylinder." Grant explained and paused, watching Dr. Morgen, who might be the big gun in this meeting.

Dr. Morgen gestured to the door and there was Amanda! She flashed a smile at Grant and took a chair beside her boss. Dr. Morgen said: "this is Amanda McCormick, my new ceramics engineer. So, please tell us about hand stoning of metal surfaces, Amanda."

She considered for a moment; she'd been unable to prepare for this meeting, Grant guessed. Her boss had probably called her when he learned of Grant's meeting.

"*Stoning* is a very old method of... burnishing the surface of hard metals, usually steels. I'm not familiar with how it would work on the aluminum-silicon composite we're using for hybrid packages. Aluminum-silicon will be a lot *softer* than any steel. I'd expect the ceramic stone would... become 'loaded-up' with the softer metal, much like what a duffer does when he tries to grind aluminum on a silicon carbide bench grinding wheel." She smiled. "I don't think we want to *smear* the surface of the composite. I'd suggest *cutting* it, if we can do that," Amanda replied.

Grant nodded. "Cutting would be both quicker and more reproducible, I'm guessing," he added, trying not to smile too much at her.

"So you want to add a final finishing step down in the cavities?" Paula asked him. He did, but he had to justify it.

"I want there to be no issues with step-like ridges left behind to cause chip or substrate *fractures*. I'd guess that either alumina or silicon carbide might be tough enough to *not* break at a stress riser. I'm worried that silicon, gallium arsenide or gallium nitride chips, being single crystalline in form, and as thin as a few mils, *might* well be cracked or cleaved... or partially cleaved by such a riser after some thermal cycling. Dr. Morgen, please counsel me; I've no actual experience with *any* of those semiconductor chip materials," Grant asked.

"Well, we don't want to *stress* a single crystal chip, especially along any crystalline axes; it might crack or cleave pretty easily," he said and paused. "You know, we *found* a failed LNA some months back where our cross-section showed just such a *step* in the package floor directly *below* the fracture in the silicon chip. I can get copies of that photomicrograph so you can all see it. Your new materials engineer is to be encouraged in trying to prevent such a failure mechanism, Paula! A few minutes per package for fine-finishing, *polishing*, might save several thousand dollars worth of diamond-filled epoxy and the LNA itself! I don't have the exact figure, but we're looking at something on the order of about $1,800 for an LNA: chips, labor, testing, laser lid-seal and final testing," Dr. Morgen replied. "That'd be 9/60 or .15 hours times something like $60 per hour: say $9.00 labor. That's trivial compared to the $1,800 package cost!"

"How would you *measure* the real, your implied word, Grant, surface finish?" Dr. Morgen asked, with just a trace of a smile.

"I'd use a surface profilometer or one of it's newer cousins: I think there's a machine called a
Dektak, that measures polished surfaces for flatness, bow, taper and smoothness, usually semiconductor wafers or chips," Grant replied, hoping he'd remembered that correctly. He needed his semiconductor process textbook!

"Very good, Grant; we have 2 or 3 Dektaks; one is up on the 4th floor of R6 in our metallurgy lab. You might enjoy getting Amanda here, to work with you on checking the surface roughness of LNA floors. Are you going to want to check before or after plating?" Dr. Morgen asked.

Grant smiled: he was being tested. "I think we could expect plating to *build up* on a ridge, making the effect worse. I think until we know more, we measure before *and* after plating. If the surface approached 1 RMS, I don't think we'd see much *decoration* of any raised feature. I think we have to be willing to invest some labor cost in learning how flat our current surfaces *are,* coming off the turning center and then how they fare in plating. We need to know what we *are* achieving and what we *might* achieve. Comments, please?" Grant asked.

Dr. Morgen leaned over and whispered something to Amanda; Grant couldn't hear what either of them said.

Paula shook her head minutely and smiled at him. She would have heard Sali describe Amanda! "I think we ask Grant to get us some typical measurements and we discuss those results. Anyone have problems with that?" Paula asked. Dr. Morgen nodded *no* to her.

"Do it, Grant. I don't want to learn we knew of an issue and ignored it! I'd like to see the photomicrograph of the cracked LNA chip, please! This is a logical addition to our core business; we make *excellent* LNAs! It would be negligent of us to ignore such an issue!" Paula requested.

They broke up to go home at the end of the work day. Grant nodded to Amanda; they would talk later about profilometry! He caught Pete's eye as the others left.

"Pete, no one found any fault with what M3 package machining does; this is a potential *improvement* in our LNA. OK?" Grant asked. He needed this guy to respect the product improvement procedure as well as his own competence.

"Yeah, I'm OK with it; Dr. Morgen is the big gun in package design... and so is Paula! We'll end up doing whatever they want!"

Grant nodded and they went their separate ways. Amanda would either walk back over here after closing up her office, or she'd call him to request he pick her up over there.

Paula shook her head, and smiled as he headed past her into his office. "Good meeting, Grant; it's a good issue to get into with exactly those people! And Amanda McCormick is a dazzler! Why isn't she a model or something?"

"Ah, I think she wants to do *engineering*!" he replied. "She is very easy to look at!" he allowed, with his own smile. Paula nodded and smiled.

About half an hour later, Amanda knocked on his office door, smiling. Grant introduced her to Leann and the two of them made quick introductory talk, as they sized each other up. Grant repressed a smile at that non-verbal, visual *assessment:* woman to woman.

A few minutes later, the two of them walked downstairs and on out to his truck. "Well, Dr. Morgen is impressed with you!" she said. He wondered how many of NG's engineers had ever considered how *smooth* the floor of a package, underneath a chip, *should* be! Good thinking, Grant!" Amanda confided.

"The Able Bond guy is the one who brought it up; he didn't say how or why he knew about it, but the machining, down in a small cavity, is fairly difficult. It would be an easy thing to overlook! I'll bet he *knows* that someone else neglected it! The epoxy is a lot softer than any metal; it *should* act as a cushion for the chip, but not if you leave a sharp edge below it! I'm sure we'll learn some more about *that,* real soon!" Grant replied with a nod.

They decided to have dinner out near Tesseract Towers, so they could commit to renting apartments. They drove out to Yorba Linda and agreed to try a restaurant across the parking lot from the apartment complex called *Manner's*. It turned out to be a quietly understated place with a fairly good menu.

Amanda ordered a steak salad; Grant ordered a small filet with a side salad; she ordered hot tea; he ordered espresso. He was pleased with how the meeting had gone; he thought Paula was pleased with how it went, but he wasn't calibrated on *any* of his new co-workers. Now he had to get some hard data on how smooth the package floors were! That part might not be so easy; he had to get the profilometer tool down in a cavity!

After dinner they walked back to the Towers and went into the sales office. Grant suggested Amanda begin the process of renting her unit. Several minutes later they had *two* salespersons assisting them! They looked at a unit on the fourth floor; Amanda did her thing of opening all the drawers in kitchen, bedroom and bath.

Grant watched, trying not to smile. When the two women salespersons understood that they would each rent *separate* 'units', hopefully not too far apart, they called up apps on their cellphones and then suggested they should move up to the fifth floor. They quickly showed 2 'units' across the hall and down from one another. Amanda and Grant inspected the two units together; both were in very good shape; both were, if not newly repainted, in excellent condition. Grant liked the one that faced north towards the golf courses of Chino Hills; Amanda liked the unit caddy-corner down the hall facing inward over the central courtyard.

After doing her thing with drawers and the closet, Amanda asked: "I guess I'll rent this one; are you OK with the other one?"

"I can be OK with it; we won't be that far apart," he replied, with a smile.

30 minutes later they'd each checked the rental agreement and signed contracts to

rent a unit for a 6-month trial period. At the end of the trial period they would have an option to renew for a standard 12 month period at a slightly reduced rate.

They took their copies of the contract agreements and went back to the truck. They would get the keys and move in the following Monday. It wouldn't be a big move to shift their things from the Residence Inn to the Towers but it would give them 'permanent' residences in California.

The next morning Grant walked over to R6 and located Amanda's office. He had a perfect excuse to work with her on his assignment: she was M & P; he would use their equipment to make the measurements necessary to determine how flat the LNA cavity-floor was. This allowed him to get the measurements and M & P could act as a 'disinterested third party'. Mechanical Manufacturing, in the person of Pete Franklin, under Don McIntyre, didn't *know* how flat their machinists were fabricating the package cavities. They might resist any change in their routine that Grant might try to implement.

Grant greeted Amanda with a smile. "So how are we going to go about this?" He asked.

"I had to read up on surface profilometry. We're going to take your test articles to one of our lab techs, Ben Warner; he's good with surface profilometers, including both the

Dektak you mentioned and the older Brown & Sharpe Tally Surf. He's inclined to use the latter instrument, if he can get the stylus down into the cavities," Amanda replied with her medium smile.

"Yeah, that may be a limit to what we can measure; these are very *small* cavities!" Grant replied.

A few minutes later, Amanda introduced Grant to Ben. Grant opened the plastic case he carried. "This production LNA is as-machined and cleaned for plating; the other one is plated with Type III gold over e-nickel over double zincate," he explained. "We should wear gloves or finger cots to keep them free of finger grease."

Wearing nitrile finger cots, Ben studied the LNAs; first with his 'Mark 1' eyeballs, and then under a stereomicroscope at up to about 40 X. "Well, there *is* some kind of machining pattern down in the cavity... floor and the plating process... either added to it or... 'decorating' it. Getting the stylus down in there is going to be a bitch! I hope we can figure out how to do that" he said, trying to visualize how to do it.

Several minutes later they had the as-machined LNA held down on a machinist's steel 'flat-plate' using small magnets to clamp the aluminum package to the steel flat. Ben gingerly fed the Tally Surf probe-arm down toward the cavity, moving the package, on the steel flat in a series of smaller and smaller adjustments to get the probe over the first cavity: the larger end 'resonating cavity', Ronnie Hartman had called it.

"Let's start with one of the larger end cavities; I know the chip doesn't go in there, according to the specs I pulled up, but I don't know whether we can *get* the stylus into that small central cavity!" Ben played with orienting the LNA on the machinist's flat to allow the stylus to enter the cavity without touching the cavity walls.

"Maybe on the diagonal, corner-to-corner?" Grant suggested. "Crossed scans might give us a good picture of how rough or smooth it is."

"Maybe. I have to *not* run the stylus into the wall, at either end! If I had an *optical* way to do this... I wouldn't have any issue. OK, I can get a trace of the center of the cavity, away from the walls, maybe," Ben decided.

He 'scanned' the stylus in the X and Y axes just *above* the bigger cavity to set the starting and ending points of the scan. Then he gently and slowly commanded the stylus to move in the Z-axis until he got a 'beep' from the control unit: the probe was now in very light contact with the conductive surface of the cavity floor. Ben swore quietly under his breath as he initiated the scan! A paper tape with an inked trace of the scanned surface began to roll out of the instrument showing a magnified view of the surface texture. "Gotta increase the mag!" he muttered under his breath.

It took him three tries to get an X-shaped scan of the larger end-cavity floor at the mechanical magnification he wanted without having the stylus strike the cavity walls. "OK, these ridges are .0019 inches above the bottom of the surface as shown by drawing a line across the bottom of the valleys. Does that make sense?" Ben asked.

"Yeah, it does; it's 'in-family' with what Pete Franklin said it would be, but that's pretty *rough*! Can you do the other big cavity, please. I don't think we can easily measure the central chip cavity, so let's argue that since the same cutting tool does all 3 cavities, the central cavity floor will be very close in roughness to the end cavities it lies between. Can I sell that?" Grant asked, looking at both Ben and Amanda.

Ben deferred to Amanda. She nodded and said: "yes, I think you can. Will Mechanical Manufacturing have *any* data they measured on their equipment?"

"I don't think so; we can ask, but I think Pete thought I was wasting his time worrying about such a... *trivial* thing. This is new to them!" Grant sighed.

"I don't think it's trivial at all; Doc Morgan told me to think about your theory that these ridges might act as stress-risers under a chip! It makes perfect sense to worry about that. Multiple thermal cycles in processing the LNA in assembly and testing, impose repeated thermomechanical stresses on the chip. I think we *know* that; I don't think that's theory! How flat, smooth, are you going to ask them for?" Ben asked.

"I don't know," Grant replied with a grin. "If they are now achieving a .002-inch surface roughness, I suppose I'd ask them to see if they can get it down to .001-inch. A one mil surface is a pretty good polish, of course. A two mil surface is not a *bad* surface, but it *might* cleave chips! Let's repeat this on the plated one. Plating might make the peaks taller, throwing 2 metals down on 2 mil peaks might produce 3 mil peaks. The semiconductor chips are only 4 mils thick!" Grant said, raising his eyebrows.

Fifteen minutes later they had scans of both end-cavities of the plated LNA. The surface finish of the plated package showed something like .0032-inch peaks and the valleys had filled in just a bit; plating liked to build-up on convex surfaces. "That's pretty much what you expected, isn't it?" Ben asked.

"Yeah, I don't know if they'll think we faked it or not!" Grant exclaimed with a chuckle. "You gotta feel good when you know how the data's going to come out!"

Grant felt pretty good; he was a novice machinist but he knew enough of the practice to make accurate guesses about surface finishes in general! Grant had the data to show Pete he had to try to achieve a smoother surface. Now he had to work with Pete to do it!

Chapter 13   Super Polish

Amanda asked: "are you gonna get all swelled-head on us?" she asked with a smile. Ben and Grant both smiled.

"Nuh-uh! Now *I* gotta figure out how to help Pete improve the surface finish effectively without jacking-up the cost very much!" Grant said with a frown.

"Isn't that *his* job?" Ben asked.

"Yeah, well... it is, but he won't work very hard on a task he thinks is trivial! Momentum usually rules! One approach would be to stone it smoother: knock off the high points. I don't know if we can expect a really smooth surface trying to *stone* them smooth by hand down in these cavities. Amanda has already suggested we *machine* them: a really *shallow* final pass. Machining might work with the right cutting tool; the part is already in a vise on the milling table. I've no idea how much such a final pass might cost," Grant said with a sigh of frustration.

"You have another option?" Ben asked, fingering the trace of the plated cavity. "The plated surface roughness approaches the thickness of our chips."

"I don't know... I was wondering about a chemical etch... well *etch-polish*, maybe. That could be done as part of the preparation for plating. A *wet* process should be capable of doing multiple parts at the same time; that

200

might be pretty cheap!" Grant said, trying to remember what he knew of chemical machining.

"An etch-polish could be batched... you might come up with a very weak acid... or base that would attack the high points and not the low," Amanda said, smiling at Grant. Grant nodded with Ben.

"That might work; we do some 'conditioning' of aluminum alloy surfaces to allow epoxy or urethane adhesives to adhere better. Maybe a cold alkaline bath? I never cleaned aluminum/silicon composites that way. That would be easy to try!" Ben suggested. Amanda and Grant nodded in unison.

"Amanda... what do you think about 'sploring' the effects of plunge-EDM... maybe *ultrasonic* plunge-EDM on Al/Si cavity floors?" Grant asked with a straight face.

'Electro-discharge machining... OK; how's *plunge* get into it?" she asked. "I thought you used a reinforced copper wire to cut shaped slots in metals? Is that not right?" she asked.

"It is, but you can also *push* a carbon or graphite rod or block into a metal: *plunge-cutting* a shaped cavity. In our case, the graphite rod would be a round-corner rectangle in cross-section, pretty much the exact size of the cavity. If the bottom, cutting surface, was *really* smooth... polished, in fact, I'd like to think the cavity bottom surface might come out pretty smooth," Grant replied, smiling.

"I gotta see this! Is *plunge cutting* in the text books? I never read much machining theory!" Amanda asked.

"We do it sometimes, over in M3, actually. It's slow or can be slow, but in this case we're just going to *touch* the machined surface. It would be pretty quick!" Ben replied, nodding to Grant.

"We're going to ask Paula if she'll pay for us to '*kiss* the surface' for a very brief time... just attack the high points! That might just work! If it does; we three might just co-author a paper: 'super-finishing Al/Si cavity floors', or something like that! That'd be a kick, wouldn't it!" Grant exclaimed, slapping the table in front of them with his fingers in excitement at his new job.

Thirty minutes later, Grant had sent an email to Paula, Dr. Morgen, Don McIntyre and Pete Franklin, showing them the data on the two kinds of surfaces from the 3 of them, using Amanda's desktop computer. He called Pete but was unable to reach him. "I'll go run him down; he's likely out on the floor. Ben, thank you! We'll come back with some more test articles, maybe tomorrow!" Grant suggested.

"You going to come learn about plunge-EDMing, *New Girl*?" Grant teased.

"Yeah, I suppose!" she replied. "You want to teach me things, I suppose?" Amanda asked when Ben was out of ear-shot.

"Well... maybe; I'm not sure about such things, you know!" he said with a smile.

"I think *that* phrase might be a 'white lie', Grant!" she replied immediately, eyes flashing.

"I think we might teach each other, actually!" he said in a whisper. She didn't respond except to smile.

"Alice, I'm going over to the M3 Machine Shop with Grant; I might be an hour or so if Dr. Morgen asks; we've sent him the results of our preliminary findings. He may want to discuss them; call me if he does."

They found Pete at his desk reading their email! "So... this is a problem?" Pete asked.

They both nodded. "We need to see if we can improve the surface smoothness of the cavity floors. There are some options about how we do that," Grant explained. He listed them: micromachining the floor with some kind of cutting tool, etch-polishing with some kind of chemical enchant, or maybe, 'kissing' the surface with a polished graphite plunge-EDM tool.

Pete considered that for a minute. "Let's go talk to Herman, he may be able to do that. Herm
Gretzky is an old Polish-German machinist. He's done a lot with EDM here at NG."

Pete found the grey-haired, weathered face machinist on break in the M3 break room, sipping coffee and reading an article in a German trade journal. Sitting at the table with the man, Grant explained what he was trying to accomplish.

Herman nodded he understood, looking at the two LNA's in Grant's sample box. "EDM... very light pressure... might clean-up that surface just a bit! You want to see what an old man can do?" he asked with a smile.

"I do! This young man and this young woman don't know *anything* of plunge EDMing except that it might remove the tops of this machined finish very quickly. We'll have to machine an electrode to match the cavity shape within about a mil or so," Grant replied.

"That's easy! I can make that electrode from a dense, fine-grained Poco graphite in half an hour!" Herm replied, still smiling.

It took Herman the rest of the day to get his regular tasks complete. At just after 5 PM, Pete called Grant and told him they'd try Herm's EDM electrode at about 8 AM the next morning. Amanda had filled-in Dr. Morgan; Grant had filled in Paula and they guessed Pete had filled-in his boss, Don McIntyre, on what they would attempt to do next.

The two of them met Pete at his desk the next morning. "Herm's got the new EDM-electrode in one of our machines; he's waiting for us. We don't EDM Al/Si composite as a general rule; turns out he played with it about a year ago and has some ideas on how he'd start doing it," Pete explained.

In the EDM room off the main machining floor, Herm showed them a simple EDM 'mill': it was an ordinary, rather old knee-mill with a stainless steel pan clamped to the X-Y table. The graphite EDM electrode was clamped in a very small vise, upside down, that gripped the graphite piece aligned in the vertical or Z-axis; power leads connected to the vise and the pan. Herm had an as-machined LNA in a vise clamped inside the pan which was filled with deionized water. The water would serve as the electrolyte that allowed the electro-discharge to occur.

Herm had three LNAs: one in the electrolyte and 2 more ready to be 'electromachined' as he called it. "Many years ago, in Gdańsk, when I was a young man, I used this process to... micro-polish some steel, *electropolishing* it. I don't know how to predict how it will work with aluminum-silicon alloy! Let's see what happens," he said his wrinkled face almost 'cracking' as he smiled.

The process was invisible; they couldn't see the surface of the LNA cavity with the electrode in place. Herm EDM'd the first LNA for 2 minutes, the second for 5 minutes and the third for 10 minutes. The cavity floor of the first LNA came out looking uniformly matt-finished: a dark gray. The 5-minute finish was light gray; the 10-minute surface was an even lighter gray. Grant thanked the man after all three LNA articles had been processed and the numbers

'2', '5' and '10' had been scribed on them so they couldn't mistake which was which.

Amanda and Grant walked them over to R6 where they waited for Ben to finish up the task he was working on. An hour later Ben clamped the '2' LNA to his surface plate: the Tally Surf scan showed a much-flatter profile: the peaks and valleys showed something about .0007-inches of 'roughness'. '5' showed something about .0003-inches of roughness. '10' showed something like .0002-inches of roughness. Laying the three new profile printouts side by side with the as-machined printout was illuminating! Herm's 'EDM-electropolish' was clearly smoothing, *flattening*, the surface of the cavity!

Amanda cut the ends of the Tally Surf scans with a scissors and placed the 4 strips of paper on a printer after labeling each of them. The scanned image showed the progressive improvement formed by what Herm had called 'electropolishing'. Pete got a copy of the progressive polishing of the LNAs to show to his people.

Amanda had done some quick research on electropolishing; it was not a new process but had been used for many years on steel, stainless steel, aluminum and titanium, using an electrolyte with the tool charged positive and the part charged negative.

Pete walked over to the Plating Shop with Amanda and Grant; Pete talked to the shop lead: Bill Miller. Pete explained what Grant was tasked to do.

"So you want me to plate these guys?" Bill asked, indicating the LNAs.

"I do; I want you to use the very same, or nearly the same process spec you normally use on LNA's. You might need to do a different *preclean* because we've EDM'd the cavity floor; I worry that we may have left some... metal smut, maybe aluminum oxide or even graphite, down in the surface texture. If you hit the LNA's with ultrasonic agitation in your normal preclean, that might be good enough. Here's how we're charging this work," and he handed Bill his charge number.

"Come back between 3 and 4; I should have them plated by then!" Bill requested.

Grant thanked Pete and Bill in advance; he and Amanda went to the main cafeteria for lunch. "Are you feeling good about how that went?" Amanda asked.

"Yeah, I am, *but!* We have to be able to deposit a smooth plating layer onto the new surface. I've read the LNA process document which details the plating procedure. The as-machined package is cleaned and double zincated; the zincate pre-plate is the key to adhesion; the third step is an electroless nickel interlayer, followed by a Type III gold plate. We need to hard solder to the gold, so the 3 layers have to be clean and adherent or it'll blister in the solder cycle at about 300 degrees C!"

"That's a pretty harsh environment; LNAs, Linear Amps and Duplexers all get the same plating on aluminum/silicon metal. If we're successful with the LNA, we might see if the process brings any improvements to the other hybrids. I'm guessing they all generate *some* heat in operation; they might benefit from higher thermal conductivity," he explained, hoping he understood the details of the procedure.

"You did some hands-on plating at Purdue?" Amanda asked.

"I did, but it was pretty simple stuff: electro-nickel over copper on different steels, followed by salt-spray testing to confirm low pinhole-count. Man, 5 or 10 percent salt-spray testing just *tears-up* steel substrates! I'd *always* use 31600 stainless if I had to subject it to ocean air! My lab instructor laughed as we watched pinholes form in marginal plating in 50 or 60 hours! 31600 stainless will go 900 plus hours before you see any rust!"

Lunch was a good break. Amanda had watched plating demos, but had never done any hands-on plating. Grant knew plating was an art! After a year in a plating shop, you were still learning basic techniques like proper deionized water rinsing!

At 3:15 PM Amanda joined Grant as they returned to the Plating Shop to find Pete, Herm and Dan McIntyre holding the plated test-articles with nitrile gloves. Pete was

beaming: the '10' LNA cavities gleamed with reflected light! The finish was very lustrous.

"Hey, Grant, Amanda! We've *never* seen surfaces this smooth! That's a sub-micron polish! If it survives the hard solder temp-cycle, I think we make both Paula and Doc Morgan happy!"

"And me; I want to be happy, *too!*" Grant said as he donned black finger cots from the dispenser on the wall and accepted the '10' LNA from Don McIntyre and studied the very reflective surface under a bright desk light.

"That's going to cost us about 10 minutes per package, but I don't think it's a bad trade-off if we get the thermal benefit you're pursuing," Don said, smiling.

Grant eyeballed the LNA and then handed it to Amanda who'd donned finger cots. "This looks almost like polished jewelry down in those cavities! This should not stress a chip mounted down there; so, now we need to make a *batch*, and get them all ready for populating, right?" she asked.

"Yes, but I think we go back and have Ben get Tally Surf scans on these 3, so our records are complete. Then we assemble feedthroughs and a thermal pad in this one to test the hard solder thermal cycle to verify there's no loss of adhesion or blistering. And *then* we can probably proceed with a batch of packages for populating and testing in M5," Grant replied, thinking about how to present the data in his report.

It took until almost noon the next day to get profile scans of the new cavity finishes. Ben handed them 3 strips of profilometer tape. Copying them onto one page gave Grant more figures that would go into his written report. He accepted Ben's assessment that Herm's 'EDM electropolish' produced a finish with about a .0003-inch peak to valley reading on the plated cavity floor. That was an *excellent* finish. By 3:30, Marti Johnson, one of Pete's solder operators, had done her post-plating cleaning and had installed feedthroughs and the copper/tungsten heat-transfer 'pad' into the '10' LNA using the hard solder: 80 gold/20 tin eutectic solder.

There was no hint of blistering! The new process was *defined*. Amanda and Grant walked the LNA over to M5 and handed it off to Jerry James for population, using some of the new SuperCon diamond-filled epoxy under the gallium nitride chip.

Grant had his report roughed out, waiting for the results of the finished LNA's electronic testing. He was pleased with the results, but he needed the final results: the LNA had to 'play' at least as well as the old units did. He hoped that Ronny Harmon would be able to report *better* performance; only that result would justify the slight increase in cost!

Paula asked Grant for a draft of his report, documenting what they'd achieved so far. He printed it out with the title: 'Draft Report: Improved LNA Finishing Procedure.' Amanda had her copy and had given a copy to Dr. Morgen.

Paula read it, smiled and scheduled a discussion at 4 PM for all the players. "It looks like you know how to do this, Grant; it looks like a textbook perfect process upgrade. When will Ronny give you the final results?"

"I don't know; she's had it in test since noon. I didn't want to hover over her, so I came back here and scrubbed the draft report. I'll go see her before 4, of course!" he replied.

"We need to update everyone so there's no question that we're fudging any data or guessing at results! *We* don't do that; OK?" Paula stated, laying down the principle of full disclosure.

"Roger that! I won't. I think Amanda will have given a copy of the draft report to Dr. Morgen, so he knows where we are. Is that acceptable?" Grant asked.

"Of course it is, but I want you to get full credit for this: it was your thorough work, and it reflects well on Manufacturing Engineering. So, your first effort here looks like a very good start, Grant. I like logical, methodical, *guided* engineering, if I can describe it that way!" she smiled. "And its none of my business, but how do you like working with Amanda?" she asked, still smiling.

Grant had expected some form of that question. "I like it; she's pretty logical. She hasn't done any plating, but undergraduate students usually don't. NG's M3 Plating Shop does excellent work! I'd like to include Jerry, Ben, Ronny and Amanda in my report as part of the technical team, if I may. That's good business!" he replied.

"It is, and you may!" she replied with a smile.

Grant walked back across the campus to M5 and entered the Electronic Test lab. Ronny was deep in a discussion with an older man Grant guessed was a EE. He stayed back, just listening. He finally figured out they were discussing the 'noise figure' of the '10' LNA coupled into Ronny's test set-up on the bench in front of her.

"These numbers are very good, Ronny! Have we ever seen anything like it?" the man asked with quiet excitement in his voice.

"*I've* never seen anything like it! The little beggar is very quiet and very cool! It's *nine* degrees C cooler than production LNA's!" Ronny said, smiling. "Oh, here he is! Dr. Williamson, this is our new materials engineer, Grant Porter."

Grant stepped forward and shook hands with the EE behind the LNA design. "Good to meet you, Dr. Williamson. So, based on a sample of one, what can we say, Ronny, about the new LNA?" Grant asked.

"We can say we see a step-function improvement in LNA hybrid performance!" Dr. Williamson interjected. "We've *never* seen numbers like this; the chip is running very smoothly and holding a low temperature under

212

load! I'm thinking about writing a paper! Tell me more about this SuperCon die-bond epoxy, Grant. Oh, I'm *Arty*, not *Dr.* Williamson, by the way."

Grant nodded. "Well, I'm speculating, but the technical sales guy for Able Bond implied we'd see cooler chip and package temps using it. I think he's supplying it to some other firms and *knows* what kind of performance it delivers. I'm really new and green, but he was very... *confident* it would work effectively. He probably can't tell us who else is using it, of course," Grant finished.

"I'll look into that! I know some guys who might be just a tad ahead of us on this! You might write a paper when we have, as you said, more than a first article!"

"Ah, I might; a process engineer's perspective of a superior thermal performance, perhaps," Grant replied with a smile.

A few minutes later, Grant hurried to make Paula's meeting which she held in M & P's small conference room. He came in just moments after the others had taken their seats.

"Late breaking news, I hope?" Paula asked, gesturing he should speak.

"Ronny Harmon and Dr. Williamson were discussing the performance of '10' LNA, the first LNA to incorporate both the smoother cavity floor and the new SuperCon die-bonding epoxy. It runs 9 degrees cooler and with an improved noise figure! I'm guessing that means it can operate on slightly less power. I

213

don't have anything in writing from Ronny yet, but I'll copy everyone as soon as I do," Grant replied. Then he gave a quick summary of what they'd done to polish the LNA cavity floors.

"*Dr.* Morgen, would you like to speculate, based on what Grant just told us?" Paula asked.

"Well, if we're going to 'Doctor' each other, *Dr.* Hopkins, I'd speculate that we ought to ask Don and Pete to see how fast we might get a *batch* of LNA's built using the new procedure; I'd expect you might start an update to the FIPP procedure based on how the majority of that batch perform. Does that sound logical?" Dr. Morgen asked as Paula smiled.

"How big a batch, Grant?" Paula asked.

"Nine would be non-trivial if they match the performance of '10'," Grant replied, not quite smiling.

"Don, are you OK with a first production batch of 9? I think we should be aggressive if we're going to find their performance superior!" Paula asked.

"I am; you better ask Grant to work with Jerry James as to how we feed them into the in-process train of hybrids," Don suggested. "I suspect we'll want to 'fast-track' them as much as we can!"

"Paula, we *may* have to redo the whole satcom design if the new LNA is that quiet! Are you ready to have Grant pursue that? Such a lower package temp may allow us to reduce power; that's a significant cost reduction and system improvement that might enable longer life on orbit!" Dr. Morgen interjected.

"Arty, you want in on this?" Paula asked as Arty Williamson joined the meeting.

"Well, I saw Ronny's preliminary results. We need to print them out... and study them some. I... I'm guessing we *rework* the Linear Amp and maybe some of the other components, as well. Grant, good job!"

"I think, with no math to back it up yet, that the cavity floor *smoothness* may be part of the *noise* in our current hybrids. If that's *true*, we need to relook at that element in *all* our packages! We're lazy: we tended to *assume* some noise was inherent. We should have studied that... in retrospect! Very interesting! I wonder what we might do if Amanda can lead us to cooler, more efficient dielectric substrates; she's indicated that we might invest some effort in SiC substrates in a couple of areas. That might help us reduce package temps another little bit! Interesting times, Paula!" Arty declared.

Paula nodded. "This information is still incomplete, but I want *nothing* we've said here to leave those in this room until Grant's report is released: copy? This is now classified *highly proprietary* until further notice: 'need to know' only. Dr. Morgen, Arty and I will determine who needs to know. Grant, you and I need to *not* tell *anyone*, including Purchasing, or Able Bond, about these results," she decided.

Chapter 14 Improved LNA

Grant wouldn't get Ronny's report until the following morning, so he waited until Amanda closed her office and the two of them walked back to M3 so he could close his office.

They decided on The Kettle for dinner; they talked in guarded terms about Grant's project during dinner. After dinner they got ice cream at the Creamery and then went back to their rooms at the Residence Inn. In Amanda's room, they were able to continue talking about the project, before they decided they needed to be sharp the next morning and retired for the night.

Grant got Ronny's data in an email the next morning at about 8:30 AM. He carefully inserted her summary in his report before printing out a copy he reread slowly. He made a copy for Paula and handed it to her in her office, waiting for her to read it. After studying it for several minutes she nodded.

"OK, Grant, let's see if we can get a production batch of 9 LNAs started. You're the lead on this; Pete knows what he needs to do; I'd let him decide on the actual tooling, but I'd try to get him thinking that maybe 'super-finishing' cavity floors is the new norm. You might want to start studying the other hybrid packages in that light. I'll get a recommendation from Arty and Dr. Morgen on what they want to upgrade next, as soon as they get copies of your report. Did we get the first shipment of SuperCon?" she asked.

"I'm supposed to get a call from M5 Stores when they have it. I have to go run that down next. Jerry will know how to do that, I'm guessing," he responded.

M5 Stores could not *find* the shipment! He enlisted Jerry's help. It took Jerry and one of his bonders to learn the shipment had been put in Receiving's freezers for safe keeping overnight, but they hadn't logged it, for some reason. Jerry got Stores to move the frozen epoxy sheets to M5 Electronic Assembly's freezer by noon. Jerry, and M5 Hybrid Assembly was now ready to build the new LNAs!

Next he had to be sure Pete's people were machining a batch of 9 LNA packages. Back in M3, Pete worked to be sure the Mechanical Fab people had the revised build documents in work. It was frustrating for Grant to work within the system to get the paperwork started before the machinists could begin cutting metal. He closed up his office Friday at just after 5:00 PM with Pete and Don's assurance that the batch of LNAs would be in work on Monday morning. He drove over to R6 and waited for Amanda to walk out.

Chapter 15   New Digs

Amanda leaned over and gave him a quick kiss once she was in the cab. "That's on account! Tesseract Towers is going to let us move in tomorrow morning! The units are empty, waiting for us and the rental agreements are complete!" she filled him in.

They went to Momma D's for dinner, having to wait out on the sidewalk for only a few minutes at the early hour. "It'll be good to get into permanent apartments won't it?" Amanda asked.

"Yeah, I'm going to take the rest of my textbooks into my office. I might refer to them there, more than I would at home," he said, thinking about that. He would enjoy settling down in his new apartment; he needed someplace to act as a semi-permanent 'home base'!

They split a rich, multi-cheese lasagna with fresh baked bread and butter. This restaurant was a bit noisy but served excellent Italian fare!

"I've got more stuff I can unpack once we lug it in from the truck! What should we do this weekend?" she asked.

"I should look for a couple shirts actually; I've been alternating a blue one with a white one, but it'd be better to have more variety. We might try walking up the Strand from Manhattan Beach to Marina Del Rey and back, if the weather's good, once we're moved-in," he suggested.

The next morning, after a quick breakfast at The Kettle, they loaded their boxes and clothes into the truck. They signed out of the Residence Inn and drove out to Tesseract Towers, where they got keys to their new apartments. They borrowed a 4-wheeled dolly and loaded the first batch of their suitcases and boxes on it. It just fit in the elevator with the two of them. They made 3 trips from the truck to the new apartments in all to get Amanda's stuff into her new unit.

Grant was pleased to have his own apartment in this very new place! New job, new friends and new digs! The newness was a bit disorienting. He enjoyed wearing jeans and a t-shirt in the Spring-like weather; the parking lot was a bit warm in the morning sun. He made a note to see what the temperature was along the coast; the Pacific ocean's temperature would temper the L. A. Basin's heat.

By noon, Amanda had fussed all her things into new places and they decided to have lunch at Farmers, across the parking lot. Grant ordered corn chowder and dark bread to go with diet cola. Amanda had a chicken salad and hot tea.

Amanda drove them back into Manhattan Beach where they found a parking space on Highland Avenue up from The Kettle.

They walked west on 13th Street to get down onto the Strand and started north on the paved lane, being passed from both directions by bike riders, skate boarders, skaters and people walking dogs.

Grant was surprised he felt some relief at having a semi-permanent place to live. He had only $180 in his checking account; he could get by until next Friday, but he felt very poor at the moment! He *needed* a first check from NG to start paying his Father back!

Walking in a nice breeze off the ocean was a new experience for him. He'd never seen anything like the Strand as they walked past the El Segundo Power Plant on the beach, watching the steady stream of people around them with the ocean as a backdrop! There were a lot of people like them just strolling in the sun, but there were a *lot* of people on bikes, skateboards and both in-line and 4-wheel skates as well. Some of the kids rode electric scooters that were faster than he'd imagined, both feet on the standing board, using just a kick now and then to go pretty fast.

"What?" Amanda asked as he studied two youngsters, a boy and a girl, maybe 8 or 9, racing down the middle of the paved strip, weaving around each other with almost no effort. The electric scooters had a fatter floor where he assumed a lithium ion battery lay.

"Those two are on *electric* scooters: they're hardly kicking at all. That's pretty fast!

I never noticed an electric powered scooter before."

A few minutes later, a pair of mid-teen girls, in bikinis, raced by on electric bicycles. He pointed to the rectangular battery attached to the seat post. "More lithium batteries! Those are e-bikes! I've read about them but never seen them! They're *moving!*" he said, smiling.

A boy, maybe 10, had a dog pulling him along on a skateboard as the boy held a leash attached to a harness; the dog was panting, but didn't seem to be exerting himself very much as they passed the two of them.

Up at Marina Del Rey, they found an ice cream shop and enjoyed waffle cones under an umbrella, watching boats moving around the marina. A group of six Penguin class dinghies raced each other in the widest part of the sheltered harbor.

"That looks like fun!" Amanda said, smiling. The sailors were mid-teen kids, maybe.

"We might rent something to sail here, sometime, Amanda! Would you like to try that?" Grant asked.

"You sail?" she asked in surprise.

"I've done some sailing in a couple of different small boats: *Lightning* and *Blue Jay* class boats," he replied.

"I watched people sail at Sandusky harbor, once, when I was younger. The yacht club set seemed to get a lot of mileage out of sailing. Where'd you sail back home?" Amanda asked.

"One of Dad's friends taught me to sail his Interlake class boat, an 18-footer up at Sandusky; I helped him strip it down and repaint it. It wasn't very fast by my lights, but it taught me the basics and some of the language of sailing. 'Believe me, my young friend, there is nothing, absolutely nothing, half so much worth doing as simply messing about in boats!'" Grant recited. "I think that's Kenneth Grahame: *The Wind in the Willows*. Small boats teach you *how* to go about doing complicated things!" he said, watching a 22-foot sloop sail back into the harbor, drop its mainsail and use the momentum and the smallish jib-sail to ease-up to the dock in its slip, precisely and easily.

Amanda took it all in, watching people, boats, and dogs on leashes as they strolled along. Grant felt very relaxed as they walked back south along the Strand and had dinner at The Kettle, before driving back out to the Towers.

"I need to settle in, somehow; I don't feel connected to this place yet. I suppose that'll take a while!" Grant confided to her as they took the elevator up to their apartments.

"I expect it will; I don't feel any connection to mine, either; I'm still a stranger in this new place. You want to be alone or do you want some company?" she asked, with a serious smile.

"I don't know," he replied. "It's only 8 and change. Let's give it an hour or so. I want to scan through some of my maps just for grins. I'll call you when I'm done with that, OK?"

※

Grant used the time to highlight some places he thought might be interesting to visit on one of his maps, now that he was a resident of Southern California. He marked some airports on the several maps; he wanted to buy a sectional aeronautical chart for the area. It would give him more details of his new home. He better buy a current California State map to annotate with places to visit, too.

He had to register his truck out here; he had to open a bank account and move his pitifully small balance out from Ohio. He'd do that on Monday or Tuesday. He needed to stop at a Macy's, maybe, and buy a couple of work-shirts. This was his first time since leaving Purdue, where he was totally on his own. It was just a bit disorienting.

It was nice to have Amanda to talk to: a kind of 'sounding board with ulterior motives'. He wished he knew what *she* wanted in a relationship; he had some fuzzy concepts of what such a relationship might be, but he didn't have a clue to what she wanted! The 'proper' thing to do would be to ask, but he didn't know if he could do that. He sat there wondering about that when there was a knock on his door!

He went to the door and looked out through the peephole. There was Amanda with a big purse over one shoulder, not quite smiling.

He unlatched and opened the door. "Hi! I was just thinking about calling you!" he said, gesturing she should come in.

She smiled; "hi yourself!" she said, leaning forward to kiss him as he closed the door and locked it. She dropped her purse on the kitchen counter and grabbed one hand, pulling him towards the 2-seat sofa, where she took a seat, still holding his hand.

She studied him at close hand for a moment before leaning forward, putting her hand on his shoulder and kissing him. It was a pretty good kiss, considering. "I like you; I think you know that. Do you?" she asked.

"I do like you; I think you know that, too! What... " she leaned in and stopped his talk with her mouth and tongue, pulling him towards her. She got up and sat on his lap, still holding the kiss, leaning against him. His body responded to her closeness, warming him up.

She ended the kiss with a sigh. "Grant, we need to have an understanding or two! You OK with that?" she asked in a whisper, her nose against his, her eyes on his at very close range.

"Yeah... I was just thinking about us, Amanda; I have no idea how to talk about this. What do you want for us, please?" he asked as she put one hand behind his head and held him there, inches from her face.

"I don't know, exactly, but I know how to start it, I think," she said, leaning her head into the angle of his neck and kissing his ear.

He responded by holding her torso against his chest, a little awkwardly on the small sofa. She turned partway towards him, pressing her breasts against him, pulling herself tightly against him. They studied each other for a few seconds before kissing again. The kisses were pretty intimate, he thought; he'd never spent this much time in an embrace.

She captured one of his hands and guided it to her breast and pressed it there. He cupped the firm breast in his hand, his pulse rocketing up: she wanted this! He started to ask something but she shook her head and just kissed him, eyes closed. He held her, one hand on her breast and the other around her shoulders. She pushed him down flat on the little sofa and lay down on him. With some adjustment, they almost fit on the sofa, feet off the end of the sofa.

He kissed her throat for a moment as she moved on him, rubbing their bodies together. She sighed as he pulled her hips against his and held her there with both hands.

"You want to undress, Mandy?" he managed to ask. She nodded, blushing. He got one hand between them and fumbled with her shirt buttons; she helped him undo some buttons. She was naked under the white shirt; his hands felt warm, smooth skin. He leaned up as she did, he captured a warm breast with his lips and tongue, causing her to sigh. She worked on his shirt with her fingers while he explored her, slowly and deliberately. She got her shirt off and his open; she pulled them together, stopping all talk by keeping her tongue in his mouth as they melted together.

He explored the small of her back with both hands, running his hand up her back to her shoulder slowly, hoping she was as excited as he was. His pulse hammered in his temples as she rubbed against him.

"I want you, Mandy! Let's shower and get into bed?" he asked. She nodded assent.

Moments later he'd closed the blinds and they stepped into the shower. He explored her body with his lips and tongue and hands as she held him by the shoulders. He soaped her slowly, enjoying every inch of her perfect body as she reacted to his exploration; then she washed him, a little more excitedly.

He handed her one of his towels and used the other to dry himself as fast as he could. In the bedroom, he pulled the blanket and sheet up so she could get into bed and then joined her, pulling her against his body, feeling her warm, smooth skin against his. He explored a perfect, firm breast with one hand while they embraced, her legs gradually opening to straddle him.

She grabbed his shoulders as their bodies moved together. His pulse pounding, he pulled her against him... her breath increased in his ear, he entered her and pulled her down on him, tighter and tighter. Her breath caught as he thrust into her, as slowly and gently as he could. He thrust up into her... she reacted against him; her eyes watched his as he penetrated deeper and deeper into her warmth, straining towards her.

"Oh," she sighed, not quite a moan, as he moved her against him, around him. He tried thrusting faster and faster; she began to breath with him, not quite panting, as they explored this very new thing together! He climaxed and died and was reborn within her. She leaned back and rubbed her breasts against him as she rode him, gently thrusting her pelvis against his, her hands on his shoulders, watching him as he recovered and began thrusting slowly up into her again. She moved with him now, more and more urgently, before she cried out and dropped down flat on him, climaxing around him.

He held her, kissing her lips and cheek as she lay there, eyes closed, recovering. After a time, she opened her eyes and with one hand around his neck, embraced him.

He lay there, her nose beside his, her eyes on his, her body full length on him, her legs around his. She smiled and studied him. "You OK, Mandy?" he managed to whisper.

She rubbed herself against him, full length, gripping him with her long legs. "Yeah... I'm very OK, Grant! *God!* I... didn't know what to expect!" she whispered. Was that good?" she asked.

"You... we were... good! I didn't know what... our bodies knew... Amazing to feel that... almost exploding! You OK with that?" he asked.

And tasted her tears! "I'm fine! After so long not knowing... now I know some of it! Oh my! That's amazing!" She pulled his head up to meet her mouth and they kissed, still entwined together. She wiped her eyes and studied him, her head turned toward his beside him on the pillow.

She rearranged herself beside him, one arm about his shoulders. "Was I... did I hurt you?" he whispered, touching her lips with one hand.

"No, not at all! The tears were something about waiting for so long! *God*, that's a powerful thing, Grant! I don't think I can control myself! Can I stay?" she asked.

"Of course *we* can! I need to see you, hold you. Is that all right?" he asked, running his hand along her back, bottom and thighs, enjoying how she reacted as he stroked her smooth skin.

"I... guessed it would be a good thing; I thought we would fit together... didn't expect it to be like *that*! My God, that was so easy, so smooth! I didn't know what to expect. Are you

OK with me?" she asked, raising her head to see his eyes on the pillow beside her.

His breathing had come down near normal; he smiled. "Yes, I'm OK with you, Amanda McCormick... we are really a *we*, aren't we?" he asked, smiling.

"We are a *we*! Oh, my God, we are a *we*!" she said smiling, and hugged him against her. She sighed as he cupped one breast in his hand and held her against him.

After a moment she put her hand over his. "You need to hold me... I'm going to sleep some. Wake me please. I'm... I'll be ready..." she said and closed her eyes, holding him with both hands.

After a while he decided she was asleep; he was afraid to move, worrying that he'd wake her. He fell asleep and woke a little later to find her looking at him, her head propped up on one hand on the pillow.

"You get any sleep?" he asked, with a grin.

"I don't need any real sleep; I need some other things, though," and she urged him up on her.

He entered her and they began again, learning more of each other as they merged again in the middle of the night. She cried out as she climaxed around him again, overwhelmed.

Chapter 16 New Relationship

In the morning, Amanda woke before Grant did; her sigh woke him up! She smiled as he opened his eyes to find her studying him again, head turned to face him on the pillow.

"Morning, sleepy-head!" she whispered.

"Did you get any sleep?" he asked, clearing his throat.

"Some! Did you?" she asked, smiling.

"I had the strangest dreams..." he tried to say before she chuckled.

"That wasn't dreaming, Grant! It was something *else*! Are you OK with *us*?" she asked, running her finger around the corner of his mouth.

"We're *committed* now, Amanda, aren't we?" he asked, unsure of what they'd started.

"I felt committed before! I decided we should go on and... have each other. I think that was a step forward. You OK with it?" she asked, hugging him with one arm across his chest.

"*Yes*, I am; I didn't see it coming until you... ambushed me out on the sofa!" he replied.

"You're very *easy* to ambush!" she chuckled, holding his torso against hers.

"Yeah, I suppose. You OK with this new arrangement?" he asked.

"I am; I did see it coming and I wanted to do that. I didn't offend you?" she asked.

"I'm not offended at all; I'm feeling some new things... I haven't sorted them out yet. Do we start talking about marriage... *hey*, are you protected from..."

"I'm on pills; Mother made sure I had a supply to bring out here. I suppose I need to get a doctor out here before too long. You OK with that?" she asked, running her finger along the line of his jaw.

He nodded. "Yes; I should have asked that last night!"

"I didn't think it was important and I didn't give you much choice, did I?" she replied. "I'm *hungry*; do you want to try to eat in or go over to Farmers?"

"We better clean up and go over to Farmers, Amanda; I don't have anything but water in the kitchen, yet! We need to go shopping; I suppose you do too. We ought to keep some kind of breakfast foods here, actually," he replied.

"Yes, we can do that; it's an extension of our new thing, *relationship*, actually! I'm your lover, I suppose; you OK with having a lover?" she asked, adjusting her body against his to watch him react.

"I am; are you?" he asked in response.

"It certainly has it's moments! Let's shower; maybe we save a little water doing it together!" she said with a smile.

An hour later they took a table near the window in Farmers. She ordered a scramble skillet to match his, with toast and orange juice. When the twin skillets arrived on wooden planks, she began eating almost immediately.

"You burned some calories last night?" he teased.

"Yeah, some; so did you!" she replied, smiling, grabbing one of his hands under the table and squeezing it.

When they were both temporarily satisfied, they decided their Sunday would involve stocking up on breakfast stuff in both apartments: oatmeal, milk, butter, eggs, orange juice and bread to toast. "We might buy some crackers and some cheese for a before-dinner snack," he suggested.

"You want to shop for some shirts?" she asked.

He nodded. "I might ask your recommendations in that. I probably tend to buy the first thing I see."

"We can go look at the Macy's down the street. They may have some good stuff. You don't want to have to iron shirts all the time. If we buy permanent press cotton, you can put on a hangar and let it dry. It'll look pretty good. You don't wear ties?" she asked.

"I don't usually. Dr. Morgan and Arty are the only ones I've met wearing ties, so far. I think I'll duck that as long as I can. I think Paula might tell me if I need a suit and tie; I don't have *any* notion as to how formal I need to be at work! You have any... notions on that?" he asked.

"You could wear a tie, but I think a clean, well-tailored pinpoint cotton oxford will go a long way, Grant! You might want to buy a little fancier pair of pants, are you likely to get any chemicals on them?" she asked.

"Uh, I don't know; if I stay out of the Plating Shop, and am careful with epoxies, I should be good. The plating baths are probably hard on clothes, but the platers all wear rubber or plastic aprons and shoe protectors. I shouldn't need to go in there very often," he guessed.

By noon, they had grocery shopped and now had minimal breakfast provisions in both fridges. Amanda got on her cellphone and located a Nordstrom's back in Hawthorne, south of Manhattan Beach. "Let's go look in this Macys and then go see what Nordstrom offers; Nordstrom will be better quality, I'm sure," she said.

By 4 PM, she had shown him more nicely tailored shirts at the Nordstrom in Galleria Shopping Center. He enjoyed squiring her around in his old truck to the very busy mall. He bought a darker blue shirt and an off-white, beige shirt, so he now had 4 shirts suitable to wear for work. "I can teach you how to iron, if you want to learn that," she offered. "They'll look a little better if you iron them. I need to buy an iron and an ironing board, actually." They stopped at a Target to buy her an iron and ironing board.

"You might wear leather shoes to work, Grant; I haven't seen enough of what the men wear. Women often wear a lot of different shoes... I don't know how to guess at men's shoes. If your leather shoes are comfortable, you might try them; I think they're a little higher on the food chain than running shoes!" she said with a smile.

She wanted to eat at McCormick and Schmick's on Marine Avenue, so they stopped for an early dinner at 5 PM. "We're celebrating *us*, OK?" she whispered to him as they were seated. He nodded, smiling his agreement.

They had a rather rich meal with a nice salad, warm, dark bread and a small fillet for him and pork rouladen for her. She drank hot tea and he drank espresso, of course.

They were back in her apartment where she set up the ironing board and the new iron, using distilled water for the steam function. The new shirts, and his old, light blue shirt, looked like new when she was done with them. She ironed a lacy white blouse she would wear with dark brown slacks.

"This is very domestic-feeling!" he declared with a smile of surprise at how good it felt. It was different but it felt right.

"It is! You OK with us being a committed couple, Grant?" she asked, eyebrows up.

"Yeah, it's growing on me, Mandy. Hey is it OK if I call you that in the privacy of our rooms?" he asked.

She chuckled. "I didn't see *that* coming! Mom used to call me that when I was little! It sounds fine, actually. I can't believe how normal, comfortable, this feels, Grant! This is very good for me!" He was there to hold her in an embrace, standing in the kitchen by the ironing board.

They spent Sunday driving out east on the 91 Freeway to Banning Pass and drove up to Mt. San Jacinto and the touristy village of Idyllwild where they meandered around, just relaxing.

✠

The next day when he went into work, he was jarred, still thinking about his new relationship with Amanda! Sali, Leann and Paula: they all had lives... were they in committed relationships, he wondered?

He met with Pete to review how they would start the 4 batches of LNA's. Grant would watch as they began the first batch of 9 silver-filled hybrid bonds, but he had to attend his first section meeting with Paula's group at 9 AM.

He thought about what he needed to do so he could become intimately familiar with how the operators did each step in the procedure. He headed upstairs to Paula's office, where she held the weekly meeting because there were only her 3 engineers and the woman who handled the group of technical writers who wrote assembly procedures, in her smallish team. When he got his iPad out of his locked drawer, he met John Lahti, his other room mate.

"I'm Grant Porter; you must be John?"

"Good to meet you! I just got back from vacation. I took my family up to Yosemite National Park; that was really nice break; we spent a couple weeks kicking around the Park. My wife had never been up there before. We rented a tent camp so it was easier for her to take care of our little girl; pretty easy camping like that! So, you're a Materials Engineering guy?" John asked.

"I am; you're a Mech-E like Leann?" Grant asked as they walked over to Paula's office. John was maybe a couple years older than he was, at a guess. Grant still wasn't sure how all the different engineers got assigned work; he'd learn more as he went.

Paula chaired the small meeting of 5 of them. She read off some status notes of her group's involvement in NG's manufacturing environment. John was involved in the support for design and manufacture of a series of standardized dish antennae: mostly carbon fiber with aluminum bases for spacecraft. Leann supported several out-of-house electronic issues working with several subcontractors. Marilyn Montose, the manager who ran the FIPP documentation group, had 5 tech writers under her direction.

"Grant, why don't you spin us up about your project!" Paula asked.

Grant spent just over five minutes explaining the thermal transfer issues his experimental assembly effort was just beginning. He would update them as soon as they had assembled the modules and had gotten them through the testing gates.

Paula added that Grant would be interfacing with M & P as his effort progressed. "I think we're going to see some parallel efforts on substrate thermal properties as our testing teaches us what works best. At the moment, another of our new engineers, one Amanda McCormick, a ceramics engineer, is heading up that effort under Dr. Morgan. Grant's work on diamond-filled film epoxy and Amanda's work on substrates may get intertwined as we learn more; polycrystalline diamond is an amazing material!"

Sali brought him a small box of 250 business cards with his name, title, the NG logo address and phone number. He tucked 4 of the new cards in his wallet and put the box in his desk. Later, he and Amanda exchanged cards at an early dinner at Darren's Restaurant in Manhattan Beach, down by the Strand. "This feels very good, Grant! It's very nice to be doing good work together! What are rock shrimp?" she asked, pointing to the menu.

"Dad recommended them to me! They're a crustacean... I think they're an Atlantic prawn; I don't know how this place gets them. Let's try them; Dad said they taste like lobster... I think they're baked instead of fried."

They shared a green salad and rock shrimp with a cup of tortilla soup. Later that evening, relaxing, back in Amanda's apartment, she studied a chapter in one of his textbooks on electronic substrates: alumina, silica, beryllia, aluminum nitride, silicon carbide and diamond.

"So, this text says it's easy to lay down a thin silicon layer by reacting silane, SiH_4, with hydrogen gas and then converting the surface of the silicon with pyrolyzed methane, CH_4 to make silicon carbide? Does that make sense to you?" she asked.

"Yeah, maybe! I wish I had more chemistry, Amanda; I have just enough to misunderstand details. But silicon carbide has been made since the early 1900's for grinding wheels and abrasives. The people who do *gas-phase* chemistry play all kinds of games with different elements. I think the pyrolysis of methane is done right below the melting point of silicon so they get some kind of reactive interface. Chemical vapor deposition, CVD, is a huge field of chemistry; a lot of our semiconductor development in the last century was based on the hydrogen reduction of several silicon chloride compounds. I think a lot of early work used germanium tetrachloride and then silicon tetrachloride. CVD chemistry is a whole... field of study! We read about it in class, and we had a... reactor, where one of the professors showed us how to grow epitaxial silicon on silicon to make diode and transistor structures. I think one can grow thin silicon layers on various dielectric materials such as alumina, quartz and beryllia. Once you get into gas phase chemistry, it's easy to make thin layers of SiC. The electronic applications just go on and on!"

She read on as Grant studied another text on chemical machining and polishing, trying to learn more about the technique Gretzky had used to produce a really flat, smooth surface.

About 9, she suggested they ought to go to bed. They showered and joined each other in her bed for the night, enjoying each other very much.

The next day, Grant dropped Amanda off at R6 and parked back by M3 where he accompanied Pete Franklin to deliver the first 9 silver epoxy LNA batch to Jerry James and Larry Elgin in M5. He would watch the M5 operators build the next batch of LNAs: the gold-filled epoxy batch. If their theory of best thermal conductivity epoxy was correct, the gold batch would outperform the baseline silver-filled epoxy. The third batch the copper-filled epoxy, was included for the sake of completeness; several of the M & P team expected that even if the copper performed well in thermal conduction, it would fail the long-term aging tests. Copper would likely react with any moisture that came into contact with aluminum, tin or arsenic in the long-term aging test. The fourth batch, the diamond-filled epoxy would likely outperform all the others.

It took another 7 days to get all 4 LNA batches completed and tested. When Ronny Harmon and 'Arty' Williamson had studied the resulting LNA performance, it was clear that diamond-filled SuperCon was a superior conductor. Dr. Morgen, Larry Elgin and Paula convened a meeting to arrive at a conclusion for the testing.

"Its pretty clear that we want to build LNA's with both the super-polished floors and SuperCon; they produce the best performance we've ever seen in low noise amplifiers!" Paula summarized. "Are you in agreement with this finding, Dr. Morgen?" she asked rhetorically.

"I am! We should fund building the PDSUb with the same technology, Paula, Grant. Do you agree with that?" Dr. Morgen asked.

"I agree. 'Arty', is that the logical way forward on this technology?" Paula asked.

"I think we could do that. We might want to build some linear integrators, LI's, as well. We may get some insights into inherent noise figure with the super-polished chip-floors, if we do that. I think we should *generalize* what we've learned and apply it across the board on some basis. Does that make sense to you, Dr. Morgen?" 'Arty' asked.

"It does. I... I think we might push our people in D1 to see if they can build SiC substrates, at least small ones for GaN chips. If we can metallize thin SiC substrates with a reliable metallization, we might see a further reduction in package temperature. 'Arty' are you knowledgable about the metallization procedure?" Dr. Morgen asked.

"I am, but it's not something we've done much with, yet! We use our in-house formula for titanium, tungsten, gold metallization, Ti, W, Au. We've two metallization systems we use for silica, beryllia and alumina substrates; it's 'old' tech! We've routinized the substrates we build in-house with Ti, W, Au for... oh, a good 4 years now. Titanium *grabs* the oxide surfaces of Si, BeO and Al_2O_3 very strongly; tungsten is a barrier layer for the gold. Gold doesn't oxidize, so we have a very strong, adherent triple layer; that means we get strong adhesion with epoxies or for eutectic bonding when we need that with the 3 older dielectric substrates. What this *doesn't* apply to, is either *SiC* or *diamond* substrates! I don't know how well Ti/W/Au bonds to SiC or the C of diamond! Isn't Amanda studying that?" 'Arty' asked.

"Yes, she is, but she's just getting started, aren't you, Amanda?" Dr. Morgan replied.

"I *am* learning, Dr. Morgen. Some of our D1 people talk of chrome/gold, Cr/Au, metallizing as an alternative to Ti/W/Au. They don't have as much 'production' history for chrome/gold, but it's a potentially simpler metallization system. I'd like to understand more of that 2-layer metallization system. We build silica, beryllia and alumina substrates in-house. We're learning how to... fabricate smallish SiC substrates, as I understand it. Isn't Wendy Jensen buying small, thin diamond substrates out-of-house?" Amanda asked. Diamond should be better than SiC, or the older 3 materials, by a *substantial* margin. I think we need to consider that as we look to improve overall thermal transduction from high-frequency RF packages," Amanda replied.

Dr. Morgan nodded agreement. "Wendy reports to William Mattison, D1 Substrates Shop Mgr. I think he determines the make/buy for substrates. I don't think we're prepared to begin trying to make *diamond* substrates; I don't know, isn't the SiC substrate process still in its infancy? We may not be able to make SiC substrates in any quantity or for an acceptable price. You want to comment on that, Amanda?"

"I believe Wendy feels they know how to build the SiC substrates within the limitations of 'small size'. I think I should defer to either Wendy or her boss, on both the maturity of the process and the costs. From what we've discussed, she thinks that CVD diamond substrates may be significantly stronger and more conductive than SiC. I don't want to try to put words in her mouth, Dr. Morgan," Amanda replied with a small smile. Grant enjoyed watching Amana join the discussion, very carefully staying out of what was another person's field of expertise.

"Grant, are you ready to assist us in getting the super-polish procedure implemented? It's easy to justify the diamond-filled SuperCon as a performance upgrade.

"I can work with Marilyn Montose to upgrade our procedures. We might want to see if we can work with SuperCon to get a price reduction or... whatever buying in more volume might do on cost. I can start that immediately, Paula!" Grant replied.

"Good! Dr. Morgen, I think we meet with "Wendy, Amanda and William Mattison in the near future. It looks like we're on the cusp of more understanding of RF microelectronics module performance!

Chapter 17   Substrates

Amanda joined Grant for a lunch in the M5 Cafeteria and introduced Wendy Jensen to him as they sat across from each other. To Grant's surprise, Ronny Harmon, walked up, carrying a tray! He introduced her to the other two women.

"Looks like an interesting place to have lunch!" Ronny said with a big smile. "I'll bet you guys are talking *diamond*!" Ronny kidded.

"Yes, we are!" Amanda replied. "We're wondering who's going to *pay* for it!" she said with a smile, as the others all smiled.

"So, Wendy and I have a meeting with Dr. Morgen and Bill Mattison at 3:30 PM to talk about substrates... maybe *advanced* substrates. I thought I could help get my head ready for that by talking about substrates for background," Amanda suggested. "Wendy, who decides which substrate gets selected?"

"Aw... that's usually an RF EE's call! If we can't use our 'old' alumina, beryllia or silica substrates because they hold back too much heat, then we're going to have to talk SiC or diamond," Wendy replied, sipping ice tea.

"So, it's a choice based on thermal efficiency?" Grant asked, wanting to know how the choice was made within the NG Redondo Beach organization.

"Mostly... cost is important, of course! We haven't made enough SiC substrates, in enough sizes, to have a really sound process or cost basis, yet. My current job is to help one technician routinize the fabrication of small SiC substrates to gain the performance improvement between the best BeO substrates and our new SiC substrates. BeO is about 330 watts per meter degree K; SiC is about 360 watts per meter per degree K... clearly superior. We've been able to get fairly good attachment using the D1 'standard' Ti/W/Au metallization procedure on our first 'standard' SiC substrate size: about .010-inches thick by about .125-inch square. That substrate works with our smaller GaAs chips and two sizes of GaN chips. We've had one batch of substrates where the metallization blistered during Ronny's long thermal post-bake at 150 degrees C," Wendy explained.

"So SiC's a little better but it's nothing like the 1,500 watts per meter degree K you'd get with a diamond substrate of the same size?" Grant asked.

"No! Several vendors are reporting *1,800* watts per meter degree K with different sources of polycrystal diamond substrates! At least one vendor reports *2,200* watts per meter degree K for their single crystal substrates, but they're very expensive!" Wendy replied. The diminutive Korean woman shook her head. "I can't see *any* path that let's us use SiC to avoid going on to polycrystal diamond! It's going to vary some depending on vendor, but PVC polycrystal diamond is 5 times better than SiC!" Wendy replied.

"We can buy... say your small substrate size from more than one vendor?" Amanda asked. Wendy nodded in silence.

"I'm just a test tech, but I'm hoping I can see an LNA built using one of these polycrystal substrates sometime soon!" Ronny added.

They talked about the various processes and materials for a little over an hour before going back to work. Grant was back in M3 working with 'Pete' Franklin and Don McIntyre as they planned how they were going to perform the super-polish on the PDSUb and the LI packages.

Don wanted to buy or build a newer, smaller plunge EDM machine to support the new requirement. Grant agreed to ask Paula to support that. When Amanda called him at 5:15 PM, he agreed to pick her up at the R6 Parking Lot. He locked up his desk and drove his pickup over to R6.

She looked tired when she got in the truck. "You look like you need some caffeine and some food, Woman!" he suggested with a smile.

"Kettle, please! Yeah, I need sustenance. Diamond is going to be expensive! Neither Dr. Morgen nor Bill Mattison can see any way around it: we're going to have to *buy* some diamond substrates to evaluate how they might help us build GaN hybrids! Wendy and I have talks scheduled to talk to two suppliers; we're chartered to buy the thinnest, least expensive, polycrystalline diamond substrates we can find. I'm afraid we're going to be looking a $75 substrates! That's a big increase over anything else we can build in house," she replied.

Grant got a Kettle-burgher, some lobster bisque and two doppio machiatto espressos as Amanda ate her burger, salad and hot tea. Afterwards they wandered down to the Creamery and got waffle-cones which they ate on a bench looking out past the Pier at the brilliant sunset off to the West.

"Wendy seemed solidly informed; how did it go with Dr. Morgan and Bill Mattison?" Grant asked.

Amanda sighed. "It went well; they're worried that we'll have to see if we can make our own SiC substrates work or buy somebody's diamond substrates! Neither of them sees anyone in-house starting up a CVD diamond substrate process! They're going to look for money to do a quick buy so we can get some preliminary test results. Good results will make it easier to find funding. Is there a chance that the guy who sells SuperCon adhesive knows somebody in diamond substrates? Dr. Morgen asked me to ask you!" Amanda asked.

"I can pulse him!" Grant agreed. "It looks like we're getting... embedded in NG's technical operation out here! I like that feeling; how 'bout you?"

"I'm feeling good about that; Ronny's interesting to talk to. Wendy likes her quick, accurate test results! Oh, Wendy and Ronny were teasing me about you! We're now 'New Girl' and 'New Guy' when we're alone, in case you wanted to know! The grapevine is very quick and accurate! Alice describes me to Dr. Morgen that way all the time! I suppose that's OK! Do you hear that?" she asked.

"Sali calls you that sometimes; I think it's just amusement. Paula uses your name; I think that's good. You still feeling tired?" he asked.

"Take me home, please, I am!" Amanda replied with a sigh.

Grant drove them East on the 91 Freeway. They showered and sat in his apartment in pajamas before turning in. It was a good end to another day in the business they were learning. Grant tucked Amanda in and stayed up, doing some planning of how he was going to finish paying his Father back for the tires he'd bought before driving out to Redondo Beach and how he was going to pay his student debt to Purdue down. It would take some months, be now he could start that!

The End

Cast of Characters

Grant D. Porter, newly minted Materials Engineer, Purdue
Mrs. Leona Miller, NG HR Manager, R4
Amy Borden, NG HR staff, R4
Robert McPage, NG HR manager, R4
Amanda McCormick, ceramics engineer, Rensselaer Polytechnic.
My-An Nguyen, mechanical engineer, Univ. of Rochester, New York
Keith Gannon, industrial engineer, USC, Los Angeles
Dan Burney, industrial engineer, UCLA Long Beach
Dr. Howard Morgen, Amanda's boss, R6
Alice Stapleton, Howard Morgen's secretary, R6
Dr. Paula Hopkins, Grant's boss, Manufacturing Eng'g M3
Sali Tompkins, Paula Hopkin's secretary, M3
Jerry James, Hybrid Ass'y tech, M5
Larry Elgin, chemical engineer, M5 under Howard Morgen
Gordon Porter, Grant's father, technical salesman
Emily Porter, Grant's mother, NP/administrator
Dr. Marilou McCormick, Amanda's mother, gynecologist
Gale McCormick, Amanda's father, city councilman, former cop
Paul Borden, IT tech

Dr. Arthur 'Arty' Williamson, EE, Spacecraft Electronic Engineering, M5
Ronny Harmon, female electrical test tech, M5
William Manchester, Able Bond field salesman
John Lahti, Mechanical Engineer, Grant's roomie, M3
Leann Davies, Mechanical Engineer, "   ", M3
Peter 'Pete' Franklin, Package Assembly Lead Tech,   M3
Annie Nguyen, Ribbon-bonding operator, M5
Marlene Eskivar, NG purchasing agent for adhesives
Harry Bond, M3 turning center machinist
Don McIntyre, M3 Mechanical Manufacturing Mgr.
Ben Warner, M&P mechanical tech R6
Herman 'Herm' Gretzky, M3 machinist
Bill Miller, M3 Plating Shop lead tech.
Marti Johnson, M3 solder tech
Marilyn Montose, M3 FIPP writer manager
Wendy Jensen, Korean D1 EE substrate researcher
William Mattison, D1 Substrates Shop Mgr.

About the Author

George Pinneo grew up in NW Ohio, graduating from Case Institute of Technology with a B.S.Ch.E. in 1959. His technical career spans 59 years of manufacturing experience, including being a founder of a successful article surveillance startup in Hollywood Florida. With 7 US Patents to his name, he has demonstrated a love of doing what had never been done before.

A recent trip took them into the Baltic region to the cities of the Hanse. Another trip took them through Andalusia in southern Spain where they stood on the top of the Rock, *Gibraltar*, and looked 8 miles south across the Straight to the mountains of Morocco rising on the North African coast near Tangiers. They enjoyed a day trip to Tangiers.

The family lives up on the Mogollon Rim of Arizona near the White Mountains. He earned a private pilot's license and built an 'experimental homebuilt' airplane, which he has spent 23 years "refining". He has twice flown cross-country to the annual EAA fly-in in Oshkosh WI. Other hobbies include woodturning, sailing, canoeing, black powder cannon and jewelry making: casting silver and bronze.

A glimpse into his notions of what constitutes good, *realistic* fiction includes solid engineering principles based on hard scientific fact, extended in a linear manner. Philosophically, good fiction should teach not only new words, but also new concepts. Good writing should be uplifting, positive and enlightening; Lois Bujold has shown that *science fiction* can also be *romantic* in her novel: "A Civil Campaign".

www.ingramcontent.com/pod-product-compliance
Lightning Source LLC
Chambersburg PA
CBHW071635220526
45469CB00002B/620